Contemporary Electrical Engineering Series

General Editors:
Professor A. H. W. Beck
Professor J. Lamb

Statistical Mechanics, Fluctuations, and Noise

2

Statistical Mechanics, Fluctuations, and Noise

A. H. W. Beck
Professor of Engineering,
Head of Electrical Engineering Division,
University of Cambridge Engineering Department

Edward Arnold

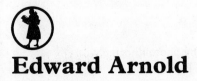

First published 1976 by
Edward Arnold (Publishers) Ltd
25 Hill Street, London W1X 8LL

ISBN 0 7131 3362 7 Boards Edition
0 7131 3363 5 Paper Edition

Text set in 10/12 pt IBM Press Roman, printed by photolithography, and bound in Great Britain at The Pitman Press, Bath

Preface

Students of electrical engineering and applied physics, especially those whose interests are in the development of electronic devices, in fusion engineering and in quantum optics, today find themselves increasingly in need of a good understanding of statistical mechanics. More especially, they require to understand and to be able to manipulate the quantum statistics of Fermi-Dirac and Bose-Einstein. Also, they require a rather sophisticated understanding of fluctuation phenomena, especially those which give rise to noise in electrical and optical communication systems. Many excellent textbooks on statistical mechanics are available, at several levels of difficulty, but few are well adapted to the needs of the audience I have described. Many are written for chemists, many more for mathematical physicists, but few try to exploit the knowledge of Fourier transforms, autocorrelation techniques and power spectral densities which engineers nowadays acquire at an early stage in their training. In this book I have tried to cover a range of statistical mechanical ideas appropriate for the understanding of semiconductor devices and of the elements of modern transport theory in gases, metals and semiconductors. At the same time, I have treated the general theory of fluctuations and the more detailed theory of electrical noise showing the link between the statistical methods of calculating the particle motions and the noise predictions of conventional circuit engineering. The mathematical techniques used in the book are based on those now available to engineers and applied physicists and I have stopped short of using the full range of techniques for the solution of stochastic differential equations, which are still being rapidly developed.

Statistical mechanics is thought of as a difficult subject and I believe that many students are put off it by too early an emphasis on the conceptual difficulties and too little emphasis on making progress to useful results. I believe that the claims made for the rigorous validity of the axioms of statistical mechanics are entirely correct and I have given some sources for this belief, but my concern has lain more with results than with postulates. One can say, without fear of contradiction, that the results usually agree well with the experiments, but it is much more satisfactory to know that the subject has a firm axiomatic foundation, than merely to appeal to experiment. I hope that many readers of the book will be interested to go more deeply into these questions.

A final chapter gives a brief outline of Hartley-Shannon information theory and the manner in which Jaynes linked it to statistical mechanics. It seems to be becoming fashionable to introduce statistical mechanics to learners by the use of information theoretical methods and enough material is included to allow my readers to understand this approach. My personal belief is that this fashion will die but that the parallel nature of the two developments remains worth description. Moreover, the bearing of information theory on the problem of measurements in noise cannot be ignored.

Most of the less advanced material has been presented in various lecture courses given to the Electrical Sciences Tripos classes at Cambridge and I am grateful to the many undergraduates who by their questions and by their answers to my questions have contributed to the presentation.

To my wife, I am indebted for the preparation of the manuscript and much other help.

A.H.W.B.

Contents

Appendices

List of Symbols

Chapter 1

P, p = probability of an event
$\mathbf{U}$ = set space
S = set space
Ω = set space
$P(B/A)$ = conditional probability of event B,
i.e. probability that B will occur if A is *known* to have done so
$f(x)$ = probability density function
$F(x)$ = cumulative distribution function
$M(s)$ = moment generating function
$C(s)$ = characteristic function

Chapter 2

s = speed of particle
u, v, w = component velocities of particle
k = Boltzmann's constant
V = volume
T = temperature, Kelvin
p = pressure
σ = collision cross section
ν = collision frequency
λ = mean free path
κ = thermal conductivity
j = current density
σ_e = electrical conductivity
D = diffusion coefficient

Chapter 3

H = Hamiltonian function
p = generalized momentum
q = generalized coordinate
$d\Gamma$ = volume element of phase space
$d\mu$ = volume element of phase space for one particle
h = Planck's constant $\hbar = h/2\pi$
w = probability of an event
$\Omega(E)$ = distribution function for the number of states
Z = partition function

β $= (kT)^{-1}$
S = entropy
F_H = Helmholtz free energy
G = Gibbs free energy

Chapter 4

g_s = number of available states
α, β = Lagrangian multipliers
E_F = Fermi energy E_{F0}, at zero temperature
θ = thermionic work function
A_0 = Richardson constant

Chapter 5

μ = Lagrangian multiplier
ν = frequency of e.m. wave.
c = velocity of e.m. wave in free space
λ = wavelength of e.m. wave
K = propagation constant $= 2\pi/\lambda$
σ = Stefan's constant
τ = Relaxation time

Chapter 6

G = torsion modulus of suspension wire
β = coefficient of friction
$\kappa(\tau)$ = autocorrelation function
$\Phi(\omega)$= spectral power density function

Chapter 7

$(\partial f/\partial t)_c$ = collision term in transport equation
C = relative velocity
$H_n(x)$ = n'th order Hermite polynomial

Chapter 8

N_p, N_e = number densities of positive ions or electrons
ω_p = 2π times plasma frequency
λ_D = Debye length
ϵ = complex dielectric constant
ϵ_0 = permittivity of vacuum
L = Coulomb logarithm

Chapter 9

R = resistance $G = 1/R$ = conductance
L = inductance
$Q = \omega L/R$
f = frequency
τ_c = time constant of circuit
η = quantum efficiency of photocathode
$F(\theta)$ = transit angle function for noise diode
P = power, P_s = signal power, P_n = noise power

Chapter 10

p = probability
H = communication entropy
C = channel (signal) capacity
F = channel bandwidth

Physical constants

Electronic charge	$e = -1.602 \times 10^{-19}$ coulomb
Electronic mass	$m = 9.108 \times 10^{-31}$ kg
$\eta = \lvert e/m \rvert$	$= 1.759 \times 10^{11}$ coulomb/kg
Velocity of light in vacuum	$= 2.998 \times 10^{8}$ m/s
Permeability of vacuum μ_o	$= 4\pi \times 10^{-7}$ henry/m
Permittivity of vacuum	$= 8.854 \times 10^{-12}$ farad/m
Planck's constant	$h = 6.625 \times 10^{-34}$ J_s
Boltzmann's constant	$k = 1.380 \times 10^{-23}$ J/K
Avogadro's number	$= 6.025 \times 10^{26}$ $kmol^{-1}$
Universal gas constant	$R = 8.317 \times 10^{3}$ J/kmol K

1

An Outline of Probability Theory

The study of statistical mechanics is impossible without some knowledge of probability theory; in this chapter we therefore outline the simpler ideas of classical probability theory. However, the use we shall make of probability in statistical mechanics differs rather sharply from the way it is used in calculating the results of coin-tossing games, card dealing, interpreting tests on samples of a population, etc., which are the meat and drink of introductions to mathematical statistics. The basic reasons for the difference, which will shortly become apparent, are that statistical mechanics deals with enormous numbers and with probability functions which are continuous functions of the independent variable. To exemplify the numbers which may be involved, we may note that typical statistical mechanical calculations on the behaviour of free electrons in metals involve the number of free electrons per unit volume, which is of the order 10^{28}–10^{29} electrons per cubic metre; thus, 10^6, which we ordinarily consider as a very large number, is quite insignificant in comparison with this density. The point about the continuous nature of the probability function is that tiresome manipulations with series are replaced by integration and differentiation. The end result will be that statistical mechanical predictions which are stochastic or probabilistic in nature turn out to be extremely precise. In general, departures from the predicted values, which are called 'fluctuations', are so small relative to the mean values that special experimental techniques are required for their detection.

1.1 The idea of probability

The definition of our meaning when we use the term 'probability of an event' has been discussed *ad infinitum* by philosophers as well as by mathematicians. It is not unfair to say that no very general agreement has been reached. For our purposes it is not necessary to go deeply into such arguments, but it is necessary to outline three ways in which probability can be defined.

1.1.1 Frequency definition of probabilities

One group of theorists (Von Mises) defines probability on the basis of the frequency with which an event occurs when a series of supposedly identical experiments is performed. For example, suppose we know that an urn contains red, white and blue counters: the experiment is to draw a counter, record its colour and return the counter to the urn. After a large number (N) of trials, the number of red counters drawn = r, of white = w, and of blue = b; then the probability on the frequency basis of drawing red = r/N, of drawing white = w/N and of drawing blue = b/N. Since we must draw either red, white or blue – because no other counters are present – we must also have the relationship

$$\frac{r}{N}+\frac{w}{N}+\frac{b}{N}=1$$

Let $\frac{r}{N} = P_{\mathrm{r}} =$ the probability of event 'red'

$$\frac{w}{N} = P_{\mathrm{w}}, \quad \frac{b}{N} = P_{\mathrm{b}}$$

Then $P_{\mathrm{r}} + P_{\mathrm{w}} + P_{\mathrm{b}} = 1.0$

In general, we shall find that the sum of the probabilities of all the possible outcomes of an experiment is unity.

As a second example of the frequency definition, suppose that we are given a penny which is known to be biased so that 'heads' are more frequent than 'tails'. If it is tossed a very large number of times and the number of heads is recorded, then the probability of heads is (number of heads) ÷ (number of tosses).

The frequency definition of probability is simple and useful because it outlines a method of measuring probability. However, it is not always appropriate and it does not, for example, give us much idea of the meaning to be associated with terms like 'very large numbers' of experiments, which have been freely used in the foregoing paragraphs.

1.1.2 Classical definition of probability

The classical definition of probability is set out by consideration of experiments as tests which have several possible outcomes or results. All the possible outcomes are assumed to be known. The probability of a specified result or event is then defined as follows:

$$\text{probability of event } x = \frac{\text{number of outcomes yielding } x}{\text{total number of possible results}}$$

For example, consider a true die (one which is carefully made to exhibit no bias). Since tossing the die can yield the six results 1 2 3 4 5 6 and no others, the probability of throwing a 6 is 1/6. The probabilities of the other possible results are all equal and have the value 1/6. Their sum

$$P_1 + P_2 + P_3 + P_4 + P_5 + P_6 = 1.0 = 6 \times 1/6$$

Notice that to apply the frequency definition we should have to throw the die a large number of times and record the results. If the large number were large enough, this would approximate 1/6 closely but would be a silly waste of time. Similarly, in the urn problem, we might well know that x red counters, y white and z blue were put into the urn; then

$$P_{\mathrm{r}} = \frac{x}{x+y+z}, \quad P_{\mathrm{w}} = \frac{y}{x+y+z}, \quad P_{\mathrm{b}} = \frac{z}{x+y+z}$$

without further experiment.

This definition clearly depends on starting with an exact description of the initial conditions. Such descriptions are rarely available in statistical mechanics; the classical definition is therefore less useful in this application than it is in the theory of gambling.

1.1.3 Ensemble definition of probability

Let us return to the idea of measuring unknown probabilities, or alternatively, of checking the predictions of theory. Suppose we wish to measure the (very small) probability that a penny, when tossed, stands upright on the thin edge rather than falling heads or tails. We could toss a particular penny a total of, for example, 10^4 times to form an estimate of the required probability. With this estimate, we could use theoretical means to consider whether the experiment needs repeating or refining. An alternative procedure is to choose 10^4 identical pennies and to place them in a machine which tosses them *simultaneously*. The number of heads, tails and edges can then be recorded. The group of identical objects is called an 'ensemble' in statistical mechanics, and the resulting probabilities–heads/10^4, tails/10^4, edges/10^4– are called 'ensemble probabilities'.

Are the ensemble probabilities identical with the frequency probabilities? Quite apart from the obvious experimental difficulties (how does one ensure that the pennies are identical and that wear does not influence the results of the frequency measurement?) there is a grave theoretical difficulty involved. It is not appropriate to delve too deeply into the problem now, but the fact is that in the early development of statistical mechanics it was thought that one could prove an 'ergodic theorem' for microscopic systems (gas molecules, etc.) which would ensure that in the course of time a microscopic system would pass through every possible state, thus making the ensemble probability equal to the frequency probability. Unluckily, the ergodic theorem is incorrect in this strong form and one can prove only that a system can pass 'near' every possible state. In spite of this problem it is always now assumed that a weaker 'ergodic hypothesis' ensures that the ensemble probabilities *are* identical with the frequency probabilities even for systems, such as pennies, which are enormously more complicated than gas molecules. It is worthwhile emphasizing here that statistical mechanics is basically an *experimental* subject and has made sharp advances precisely when the existing theory did *not* correctly predict the results of some experiments. A familiar example of this is the failure of early statistical mechanics to derive useful results for the thermal and electrical conductivities of metals.

We have now three definitions of probability. In statistical mechanics, the frequency and ensemble definitions are the important ones and we shall use both freely. From the mathematical viewpoint it is unsatisfactory not to start from a unique definition which would allow probability theory to be based on a purely axiomatic foundation, but as physicists we are faced with other problems and the flexibility gained is helpful in solving these.

1.2 The axiomatic basis of modern probability theory

Modern mathematical probability theory is based on the assertion of a series of axioms which are not themselves capable of proof by the theory developed from them. Frequently the classical definition of probability is taken as the starting point. In his reformulation of probability theory, the Russian mathematician Kolmogorov relates this theory to the theory of sets, but we shall require only to introduce a small amount of terminology from set theory.

A set space (symbol **U**, S or Ω) is defined as comprising the elementary events associated with a particular experiment. Certain sub-sets F of **U** are called 'random events'.

F has the following properties.

1. F includes **U** as one of its elements.

2. If the sub-sets A and B of **U** are included in F, then so are $A + B$ (often written $A \cup B$ and meaning those elements which are in A or B or both), AB (often written $A \cap B$, and meaning those elements which are in *both* A and B), $\bar{A}$ (the complement of A, i.e. those elements *not* in A) and $\bar{B}$. Thus, F includes **U** or V, the so-called 'empty set'. This follows because **U** is a member of F.

3. If the sub-sets $A_1, A_2, \ldots, A_n$ of set **U** are elements of F, so are the *sum* $A_1 + A_2 + A_3 + A_n + \ldots$ and the product $A_1 A_2 A_3 \ldots A_n$.

F is then termed a 'closed field of events'.

The axioms are now asserted as follows.

First axiom. With each random event in a field of events there is associated a non-negative number, called its 'probability'. We use the symbols $P(A)$, $P(B)$ etc.

Second axiom. $P(\mathbf{U}) = 1$

Third axiom. If the events A_1, A_2, A_3 are pairwise mutually exclusive, we have $P(A_1 + A_2 + A_n) = P(A_1) + P(A_2) + P(A_n)$ etc. Mutually exclusive events are, of course, those that *cannot* occur together; for example, the events of throwing the several faces of a die are mutually exclusive. (This axiom is the law of addition of probabilities.)

The following consequences of the axioms are obvious.

1. Since $\mathbf{U} = \mathbf{V} + \mathbf{U}$, the third axiom immediately says that $P(\mathbf{U}) = P(\mathbf{V}) + P(\mathbf{U})$ and therefore the probability of the empty set (the impossible event) is zero.

2. Since $\mathbf{U} = A + \bar{A}$,

$$P(\bar{A}) = 1 - P(A)$$

3. $0 \leq P(A) \leq 1$

To proceed further, it is useful to introduce the Venn diagram. In this diagram the area **U** denotes the sample space and areas A and B represent the probabilities of outcomes A and B respectively. In Fig. 1.1(a) we represent $A + B$ or $A \cup B$; in Fig. 1.1(b) the product $AB (A \cap B)$ and the complement $\bar{A}$. In Fig. 1.1(a) the areas A and B are drawn to overlap; this means that the events are *not* mutually exclusive but, instead, there is a finite probability of both occuring.

However, from Fig. 1.1(a) we can see geometrically that

$$P(A + B) = P(A) + P(B) - P(AB)$$

This follows because $P(A) + P(B)$ clearly includes *twice* the area representing $P(AB)$. Since $P(AB)$ is non-negative, it also follows that

$$P(A + B) \leq P(A) + P(B)$$

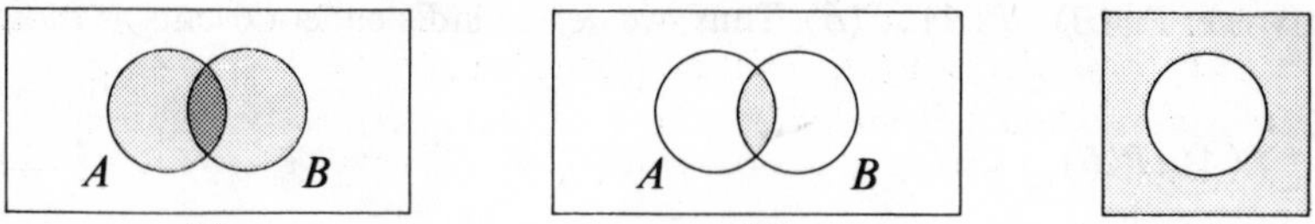

(a) Shaded area is union (sum) of A and B: symbol $A \cup B$

(b) Shaded area is intersection of A and B: symbol $A \cap B$

(c) Shaded area is complement of A: symbol $\bar{A}$

Fig. 1.1 Venn diagrams, illustrating (a) the union or sum of A and B, (b) the intersection of A and B, (c) the complement of A ($\bar{A}$).

Venn diagrams are of very great use in visualizing the theorems of discrete probability.

Another important axiom is easily illustrated by the Venn diagram. This is the axiom of addition, stating that, if event A is equivalent to at least one of the pair-wise mutually exclusive events A_1, A_2, A_3, etc., then

$$P(A) = P(A_1) + P(A_2) + P(A_3) + \ldots$$

1.3 Conditional probability

The concept of conditional probability is important in statistical mechanics. The conditional probability is defined through the idea of two tests in sequence, and the conditional probability $P(B/A)$ is the probability that event B will occur *if event A is known to have occurred.* Thus, if we make two draws from a pack of cards, we might wish to know the conditional probability of drawing a second ace when the first trial was known to be an ace. Clearly, after the first trial there are three aces left among 51 cards. Therefore, in this case, the conditional probability is 3/51. In non-equilibrium statistical mechanics we frequently wish to find the probability that a particle initially at $z = 0$, $t = t_0$ is at $z + \Delta z$ at $t_0 + \Delta t$, or three-dimensional variants of the same problem. This probability is a conditional probability according to our definition.

If we return to the relevant Venn diagram, Fig. 1.1(b), we see that the shaded area represents the area in which B can occur, A having certainly happened. Thus, the formal definition of $P(B/A)$ is

$$P(B/A) = P(AB)/P(A) \qquad 1.1$$

similarly, $P(A/B) = P(AB)/P(B)$ 1.2

and, in general, these are not equal. Combining equations 1.1 and 1.2 and we obtain the important multiplication theorem

$$P(AB) = P(B/A)\,P(A) = P(A/B)\,P(B) \qquad 1.3$$

Consider now what happens if the two events A and B are *stochastically independent,* that is if knowledge of the outcome of one test has no influence on the probability of the other. This means that the conditional probability $P(B/A)$ should be identical with $P(B)$, and $P(A/B)$ should equal $P(A)$. From equations 1.1 and 1.2

that this involves $P(AB) = P(A) . P(B)$. Thus, we *define* independent events as those for which

$$P(AB) = P(A) . P(B) \tag{1.4}$$

In physics, we often do not know whether events are really independent or not. What is usually done is to *assume* that they are independent and to compute $P(AB)$ from equation 1.4, which is therefore an outstandingly important relationship. It is sometimes called the 'multiplication law'.

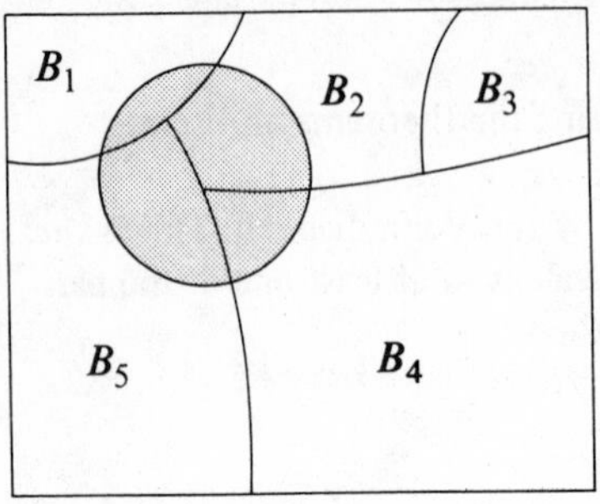

Fig. 1.2 Venn diagram illustrating the multiplication law. A is the shaded area.

Let us now generalize the multiplication law. Consider Fig. 1.2, which is the Venn diagram for a case in which there is an event A which depends on several events $B_1 \ldots B_n$. However, the events B are mutually exclusive and exhaust the possible outcomes of the test, as shown by the fact that their representative areas do not overlap. From the geometry of the diagram we can see that

$$P(A) = P(AB_1) + P(AB_2) + \ldots + P(AB_n) = \sum_{i=1}^{n} P(AB_i) \tag{1.5}$$

But, using equation 1.3 repeatedly,

$$P(A) = P(A/B_1)\, P(B_1) + \ldots + P(AB_n)\, P(B_n)$$

$$= \sum_{i=1}^{n} P(A/B_i) . P(B_i) \tag{1.6}$$

This result is not affected by the fact that in our Venn diagram $P(A/B_3) = 0$, since A excludes B_3.

1.4 Continuous random variables

So far, the probabilities of the events we have considered are discrete numbers whose sum is unity. In many cases of interest in physics and engineering $P(x)$, the probability of the event x, is a continuous function of x. As a very simple example, consider an experiment in which a thin needle is tossed so that it intersects a line of known length L at a point x, whose coordinate can be accurately measured along the line. We can represent the situation by Fig 1.3, since the total probability of all possible events, which is represented by the shaded area, must equal unity. Now, notice that the point probability $P(x)$ has lost its simple meaning. Instead, the diagram conveys the information that the probability of finding the needle between x_i and $x_i + \Delta x$ is the double-hatched area, equal to $\Delta x/L$. This probability $\rightarrow 0$ as $x \rightarrow 0$, so that the point probability is zero. This situation is more unsatisfactory in classical physics than it is in quantum physics, where the uncertainty principle

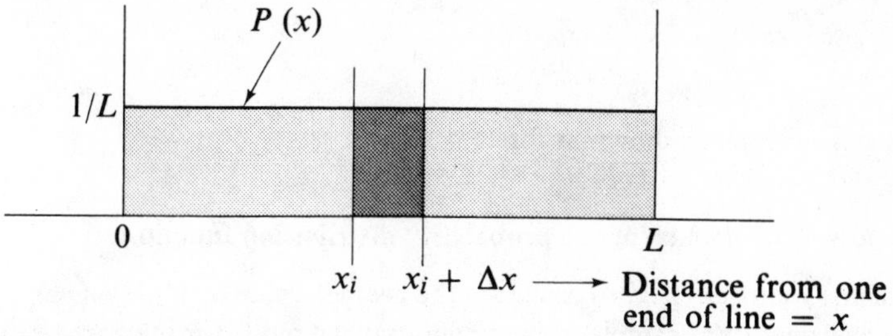

Fig. 1.3 The probability distribution for a random point on a line length L.

specifies a minimum value for Δx and therefore a minimum probability for the location of the needle.

Thus we come to the following *definition.* Let x be a continuous random variable. The *probability density function* (p.d.f.), $f(x)$, associated with x must satisfy the following conditions:

1. $f(x) \geqslant 0$ for all x in the range of x, $\{R(x)\}$

2. $\int_{R(x)} f(x)\,dx = 1$

The probability of finding a value of x between a and b, $b > a$, is

$$P(a < x < b) = \int_a^b dx$$

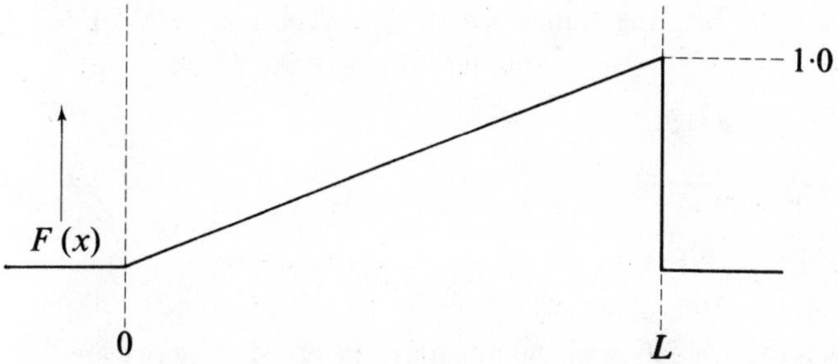

Fig. 1.4 The cumulative probability that the random point will lie in the range $0 \rightarrow x$.

It hardly needs to be said that $f(x)$ can be any continuous function or it may have discontinuities. Often, the range of x is from $-\infty$ to $+\infty$ or, if it is not but the probability density is zero outside limits c and d say, it may be convenient to extend the range of integration from $-\infty$ to $+\infty$, to utilize the techniques of Fourier transformation.

In making probabilistic calculations, the co-called '*cumulative distribution function*' (c.d.f.) is also frequently required. This gives the probability of values of x up to and including some specified value. The c.d.f. (symbol $F(x)$) is defined by

$$F(x) = \int_{-\infty}^{x} f(\zeta)\,\mathrm{d}\zeta$$

The c.d.f. for our example is shown in Fig. 1.4.

1.5 Averages of powers of x taken over a probability distribution function

In physical problems we often need to calculate the average value of some power of the independent variable. For example, it is known that the p.d.f. for the speed of a molecule of an ideal gas varies as $s^2 \exp(-\alpha s^2)$; we might wish to compute the mean kinetic energy of a molecule. The number of particles in the (very small) range of speeds, s_1 to $s_1 + \Delta s$ is $ks_1{}^2 \exp(-\alpha s_1{}^2)\Delta s$. We usually express this by saying that it is the number in Δs at s_1. Then, assuming we know that there are N particles in all, we can determine k from

$$N = \int_{-\infty}^{+\infty} ks^2 \exp(-\alpha s^2)\mathrm{d}s \qquad 1.7$$

where we have assumed that the s values are so close together that we can use integration instead of sums.

The group of particles in Δs at s_1 makes a contribution to the kinetic energy which is given by $\frac{1}{2}ms_1{}^2 \,.\, ks_1{}^2 \exp(-\alpha s_1{}^2)\Delta s$.

Then, the total k.e. is

$$E_{\mathrm{T}} = \int_{-\infty}^{+\infty} \tfrac{1}{2}ms^4\, k\,.\,\exp(-\alpha s^2)\mathrm{d}s \qquad 1.8$$

By definition, the *average* k.e., the average k.e. of one particle, is E_{T}/N, or

$$\langle E_{\mathrm{T1}}\rangle = \frac{\int_{-\infty}^{+\infty} \frac{1}{2}mks^4 \exp(-\alpha s^2)\,\mathrm{d}s}{\int_{-\infty}^{+\infty} ks^2 \exp(-\alpha s^2)\mathrm{d}s} \qquad 1.9$$

We shall later discuss the evaluation of the integrals concerned: here, we are interested only in the method. This can be generalized as follows.

Let $f(x)$ = p.d.f. of x, and suppose that we wish to find the average of some function $g(x)$ of x. Then we can write

$$\langle g(x)\rangle = \frac{\int_{-\infty}^{+\infty} g(x)f(x)\,dx}{\int_{-\infty}^{+\infty} f(x)\,dx} \qquad 1.10$$

But, since $f(x)$ is a properly normalized probability function, the lower integral is by definition unity and we have

$$\langle g(x)\rangle = \int_{-\infty}^{+\infty} g(x)f(x)\,dx \qquad 1.11$$

The beginner is well advised not to use this last form too readily, otherwise he will make mistakes when dealing with distribution functions which have *not* been properly normalized. As another hint, it is always useful to remember that if an integral which is of the form $\int_{-\infty}^{+\infty} g(x)f(x)\,dx$, where $f(x)$ is a distribution function, crops up in a calculation, it is merely an expression for the average of $g(x)$. Because the average is often known from some outside information, it is unnecessary to waste time on working out elaborate integrations. For example, the average k.e. per molecule of the ideal gas is $3kT/2$, where k = Boltzmann's constant. Evaluation of the defining integrals is not required, except in examination questions!

In conclusion, it should be noticed that, if the probabilities are discrete, the above formulae must be altered by replacing the integrations by summations.

1.6 The moment-generating function and the characteristic function

Continuing with the ideas of the last section, it frequently happens that we know the values of $\langle x\rangle$, $\langle x^2\rangle$, $\langle x^3\rangle$, etc. without knowing the exact mathematical representation of the p.d.f. $f(x)$. The relationships to be derived in this section show how we can derive the p.d.f. (or a useful approximation to it) from a knowledge of all, or some, of the averages. The first, less useful, of these techniques uses the *moment-generating function.* This is defined by

$$M(s) = \int_{-\infty}^{+\infty} \exp sx\, f(x)\,dx \qquad 1.12$$

Here, s is purely real and $f(x)$ is a p.d.f. Notice that, by definition,

$M(s)$ = average of $\exp sx$

Now, the absolutely convergent series which represents exp sx is

$$\exp sx = 1 + sx + \frac{(sx)^2}{2!} + \frac{(sx)^3}{3!} + \ldots + \frac{(sx)^n}{n!}$$

therefore

$$M(s) = \langle \exp sx \rangle = \left\langle \left(1 + sx + \ldots . + \frac{(sx)^n}{n!} + \ldots\right) \right\rangle$$

Proceeding very formally, without justifying the processes, we can write

$$M(s) = 1 + s\langle x \rangle + \frac{s^2 \langle x^2 \rangle}{2!} + \ldots + \frac{s^n \langle x^n \rangle}{n!}$$

since s is merely a constant parameter. Differentiating this with respect to s, we obtain

$$\frac{\mathrm{d}M(s)}{\mathrm{d}s} = \langle x \rangle + \frac{2s\langle x^2 \rangle}{2!} + \ldots + \frac{s^{n-1}\langle x^n \rangle}{(n-1)} \qquad 1.13$$

and, putting $s = 0$, then

$$\left[\frac{\mathrm{d}M(s)}{\mathrm{d}s}\right]_0 = \langle x \rangle \qquad 1.14$$

Similarly, $$\left[\frac{\mathrm{d}^2 M(s)}{\mathrm{d}s^2}\right]_0 = \langle x^2 \rangle \qquad 1.15$$

$$\left[\frac{\mathrm{d}^3 M(s)}{\mathrm{d}s^3}\right]_0 = \langle x^3 \rangle \qquad 1.16$$

From the analogy with moments of inertia, etc., statisticians call the averages $\langle x \rangle$, $\langle x^2 \rangle$, $\langle x^3 \rangle$, the first, second, third moments of the distribution, and these last relationships show why they speak of the 'moment-generating' function $M(s)$.

Rather than considering examples of the use of $M(s)$, we go straight on to the *characteristic function*. This is *defined* by the integral

$$C(s) = \int_{-\infty}^{+\infty} \exp(\mathrm{j}\,sx)\,.\,f(x)\,\mathrm{d}x \qquad 1.17$$

The characteristic function will thus be recognized immediately as the Fourier transform of the p.d.f. We can therefore use all the elaborations of Fourier-transform theory without further thought, because by definition the integrals over $f(x)$ are absolutely convergent. The Fourier mate of $C(s)$ is also immediately given by

$$f(x) = \frac{1}{2\pi} \int_{-\infty}^{+\infty} C(s) \exp(-\mathrm{j}\,sx)\,\mathrm{d}s \qquad 1.18$$

As before,

$$C(s) = \langle \exp(\mathrm{j}sx) \rangle \qquad 1.19$$

Therefore $C(0) = 1.0$ 1.20

If we again expand $C(s)$ as a MacLaurin series:

$$C(s) = 1 + \sum_{n=1}^{\infty} \frac{(\mathrm{j}s)^{\mathrm{n}}}{n!} m_{\mathrm{n}} \tag{1.21}$$

then

$$m_{\mathrm{n}} = \frac{1}{\mathrm{j}^{\mathrm{n}}} \left| \frac{\mathrm{d}^{\mathrm{n}}\, C(s)}{\mathrm{d}s^{\mathrm{n}}} \right|_{s=0} \tag{1.22}$$

Sometimes it is more convenient to use a set of quantities called the 'cumulants' rather than the moments. These are defined through the relationship derived by expanding ln $C(s)$ about $s = 0$:

$$C(s) = \exp\left(\sum_{n=1}^{\infty} \frac{(\mathrm{j}s)^{\mathrm{n}}}{n!} k_{\mathrm{n}} \right) \tag{1.23}$$

The cumulants are related to the moments by

$k_1 = m_1$
$k_2 = m_2 - m_1{}^2$
$k_3 = m_3 - 3m_1 m_2 + 2m_1{}^3$
etc.

Thus, k_1 is the mean of the p.d.f.
Let us examine k_2.

$$k_2 = \langle f^2(x) \rangle - \langle f(x) \rangle^2 \tag{1.24}$$

and thus equals the variance or dispersion of the p.d.f. The smaller the value of k_2, the more sharply peaked is the p.d.f. Finally, the standard deviation $\sigma = \sqrt{k_2}$ is given by

$$\sigma(x) = [\langle f^2(x) \rangle - \langle f(x) \rangle^2]^{1/2} \tag{1.25}$$

therefore, σ has the same dimensional form as the independent variable.

1.7 The binomial distribution

We next turn our attention to the problem of the expected results of a series of experiments when each experiment has only two possible outcomes, symbolically A and $\bar{A}$, with probabilities p and q respectively ($q = 1 - p$). The best-known example is that of tossing a coin, in which case $p = \frac{1}{2} = q$. Exactly the same theory applies to the more interesting calculations involved in evaluating the magnetic moment of a system of electrons, which have spins of $+\frac{1}{2}$ and $-\frac{1}{2}$. An external magnetic field here modifies the probabilities so that more spins will be found in one sense than in the other.

For the moment, we shall concentrate on coin tossing, which lends itself to a degree of experimental verification without requiring complicated apparatus, merely patience. Suppose we toss the coin N times and count the numbers of heads (probability p) and tails (probability q) and find n heads and $N-n$ tails. What is the probability of this particular result?

Let $\binom{N}{n}$ = number of distinguishable configurations in which n heads occur.

Then, clearly the *probability* of observing this is

$$P(n) = \binom{N}{n} p^{n} q^{N-n}$$

How does one evaluate $\binom{N}{n}$? This is a problem of combinatorial mathematics, and the argument starts by observing that the number of ways in which n objects can be chosen from a total of N is

$$N(N-1)(N-2)\ldots(N-n+1) = \frac{N!}{(N-n)!}$$

However, this is not the number of *distinguishable* arrangements or configurations. We can permute the n objects among themselves without changing the essential observation that we wish merely to achieve n heads; thus, on division by $n!$, which is the permutation of n identical objects,

$$\binom{N}{n} = \frac{N!}{n!\,(N-n)!}$$

$\binom{N}{n}$ is often called the 'number of combinations' for this choice of n from N.

Another symbol, $C_N(n)$, is frequently found in the literature. Writing out $P(n)$ explicitly, we find

$$P(n) = \frac{N!}{n!\,(N-n)!}\, p^{n} q^{N-n}$$

When $p = q$, which obviously happens only when they both equal $\frac{1}{2}$, we find

$$P(n) = \frac{N!}{n!\,(N-n)!}\left(\frac{1}{2}\right)^{N}$$

Comparing these results with the binomial expansion of $(p+q)^N$ we see that

$$(p+q)^{N} = \sum_{n=0}^{N} \frac{N!}{n!\,(N-n)!}\, p^{n} q^{N-n} = \sum_{n=0}^{N} P(n)$$

which explains why our result is called the 'binomial' probability distribution. It also makes directly obvious the result that

$$1 = \sum_{n=0}^{N} P(n)$$

so the $P(n)$ are properly defined probabilities.

1.7.1 The mean, or average, value of the binomial distribution

By definition, the average value of the number of heads obtained in n tosses is

$$\langle n \rangle = \sum_{n=0}^{N} n\, P(n)$$

$$= \sum_{n=1}^{N} \frac{N!}{(n-1)!\,(N-n)!} p^{n} q^{N-n}$$

Putting $\alpha = (n-1)$, we obtain

$$\langle n \rangle = \sum_{\alpha=0}^{N-1} \frac{N!}{\alpha!\,(N-1-\alpha)!} p^{\alpha+1} q^{N-\alpha-1}$$

$$= Np \sum_{\alpha=0}^{N-1} \frac{(N-1)!}{\alpha!\,(N-1-\alpha)!} p^{\alpha} q^{N-\alpha-1}$$

$$= Np\ [(p+q)^{N-1}] = Np \qquad 1.26$$

The last relationship follows since $p + q = 1$. The mean or average is often called the 'expectation value' in statistics, and what we have symbolized as $\langle n \rangle$ is written $E(n)$. It seems unnecessary to introduce this extra term, although it is clear that if we make a large number of tosses we can 'expect' Np heads. Unfortunately, if we toss a coin 10^3 times and obtain 455 heads, i.e. rather less than 500, we cannot conclude that the coin is dishonest without a much more far-reaching investigation.

Using a minor modification of the method used to evaluate $\langle n \rangle$, it can be proved that

$$\langle n^2 \rangle = Npq + N^2 p^2$$

for the binomial distribution. The trick is to write

$$\frac{nN!}{(n-1)!} = \frac{(n-1)\,N!}{(n-1)!} + \frac{N!}{(n-1)!} = \frac{N!}{(n-2)!} + \frac{N!}{(n-1)!}$$

As we have said earlier, a more useful parameter of the distribution is the mean square of the deviation from the mean value. This quantity is called the '*variance*' of the distribution. The variance is given by

$$\begin{aligned}\langle(n-\langle n\rangle)^2\rangle &= \langle(n^2-2n\langle n\rangle+\langle n\rangle^2)\rangle \\ &= \langle n^2\rangle - \langle n\rangle^2 \\ &= (Npq+N^2p^2)-N^2p^2 \\ &= Npq \qquad 1.27\end{aligned}$$

The square root of the variance is dimensionally the same as the n or $\langle n\rangle$ and is called the 'standard deviation' of the distribution. It is often given the symbol σ. For the binomial distribution,

$$\sigma = \sqrt{(Npq)} \qquad 1.28$$

The 'width' of the distribution can be measured by comparing σ with $\langle n\rangle$; then

$$\frac{\sigma}{\langle n\rangle} = \sqrt{\frac{q}{Np}} \qquad 1.29$$

Thus, the larger N the sharper the peak in the distribution function, independently of the values of q and p.

The binomial distribution has just been shown to have means and variances given by extremely simple formulae, but if one attempts to compute the actual probability of a particular set of outcomes from a large number of tests, it is at once obvious that handling the factorials is a tiresome and inaccurate process. We therefore next discuss the approximation of the binomial distribution by analytical functions.

1.8 The Poisson distribution

This approximation is valid under the following conditions:

1. $N \gg 1$
2. $p \ll 1$
3. $n \ll N$, but np is finite

A rough and ready derivation of the Poisson distribution formula is as follows. We can approximate $\dfrac{N!}{(N-n)!}$ by $N(N-1)(N-2)\ldots(N-n+1)$ which, because $n \ll N$, becomes

$$\frac{N!}{(N-n)!} = N^n$$

Also, $q^{N-n} = (1-p)^{N-n}$ can be simplified.

Taking logs, we have $\ln q = \ln(1-p)$

But, if $p \to 0$, $\ln(1-p) \to -p$ and $\ln q \to -p$.

or $\quad q = \exp(-p)$

$$\begin{aligned}q^{N-n} &= \exp -p(N-n) \\ &\approx \exp -pN\end{aligned}$$

With these result, the Poisson distribution becomes

$$P(n) \approx \frac{(pN)^n}{n!} \exp(-pN) \qquad 1.30$$

Since n is supposed fairly small, it is not difficult to evaluate $n!$ This formula is often written with $\lambda = pN$ inserted. It becomes

$$P(n) = \frac{\lambda^n}{n!} \exp(-\lambda) \qquad 1.31$$

λ is, of course, the mean value of the original binomial distribution. The above derivation can be accomplished by much more accurate and elegant means and the result is, in fact, an excellent approximation under the cited conditions. However, let us proceed to establish that the average taken over this distribution is, in fact, pN, as it ought to be if the binomial distribution is accurately approximated.

$$\text{By definition, } \langle n \rangle = \sum_{n=0}^{\infty} \frac{n\lambda^n}{n!} \exp(-\lambda)$$

$$= \lambda e^{-\lambda} \sum_{n=0}^{\infty} \frac{\lambda^{n-1}}{(n-1)!}$$

$$= \lambda e^{-\lambda} \sum_{\alpha=0}^{\infty} \frac{\lambda^{\alpha}}{\alpha!} = \lambda e^{-\lambda} . e^{\lambda}$$

$$\text{Therefore} \quad \langle n \rangle = \lambda = Np \qquad 1.32$$

In similar fashion, it can be proved that $\langle n^2 \rangle = \lambda^2 + \lambda$. Thus the variance is λ and the standard deviation $\lambda^{1/2}$. In this case the ratio

$$\frac{\sigma}{\langle n \rangle} = \lambda^{-1/2} = (Np)^{-1/2} \qquad 1.33$$

Thus, the Poisson distribution has the same mean as the binomial distribution but the variance is only approximately the same, as it is derived from the binomial variance by letting $q \to 1$. Since

$$\sum_{n=0}^{\infty} \frac{\lambda^n}{n!} \exp(-\lambda) = 1$$

the Poisson distribution is a properly normalized probability distribution.

So far our derivation has tacitly assumed that n is restricted to integral values, so that the Poisson distribution is being thought of as a discrete distribution. However, if we define $n!$ through the gamma function, we see that

$$n! = \Gamma(n+1) = \int_0^{\infty} z^n \exp(-z)\,dz$$

where z is a dummy variable. Thus, $n!$ is defined for all real n and is not restricted to integral values. The Poisson distribution can therefore be regarded as a continuous analytic function and the tedious summations we have used can be replaced by simple integration.

At this point, it is important to realize that, although we started out from a somewhat rough approximation to the binomial probability model, we have finished with a perfectly proper mathematical function which describes a probability strictly according to the axiomatic foundation of the theory. It is therefore natural to enquire if there are physical systems or processes which can be described by the Poisson distribution without any necessity for proceeding through the binomial model. The answer is affirmative, and there are, in fact, many processes which are adequately described in this way. Examples are the number of electrons emitted from a thermionic cathode, and the closely allied problem of radioactive emission. Poisson distributions are also found to describe the number of telephone calls entering an exchange per unit time, the number of customers arriving at a pay desk etc.

1.9 The Gaussian or normal distribution

If the binomial distribution is plotted for even quite small values of N and for p and q comparable in magnitudes, the shape of the resulting histogram strongly suggests that the envelope curve should be capable of approximation by a Gaussian curve, that is the function constant x $\exp(-\alpha x^2)$ about the mean value of the binomial distribution. This conjecture is correct and the process will again yield a mathematically proper, continuous, probability distribution. The Gaussian distribution is perhaps the most important distribution function in the whole of probability theory and statistical mechanics. For this reason, we include a fairly complete account of its properties.

The passage from the binomial distribution to the Gaussian has to be carefully studied in statistics dealing with relatively small values of N, and the nature of the approximations made is important in this case; in statistical mechanics the number N is usually enormous and questions about accuracy of approximation do not arise. The passage is achieved by expanding the several factorials involved using Stirling's approximation, or, more strictly, one of Stirling's approximations, for they form a heirarchy of increasing accuracy for small arguments. First, we write the binomial probability in a slightly modified form, introducing the deviations of n from $\langle n\rangle = Np$.

Let $n = Np + u$

then $(N-n) = N(1-p) - u = Nq - u$

and $$P(u) = \frac{N!}{(Np+u)!\,(Nq-u)!}\,p^{Np+u}\,q^{Nq-u}$$

Taking the logarithm of this gives

$$\ln(P) = \ln N! - \ln(Np+u)! - \ln(Nq-u)! + (Np+u)\ln p + (Nq-u)\ln q$$

The form of Stirling's approximation we use is

$$\ln x! = x\ln x - x + \tfrac{1}{2}\ln 2\pi x^*$$

If this is used to expand all the factorials and if we expand quantities such as $\ln(Np+u) = \ln Np + \ln(1+u/Np)$ in series, getting

$$\ln(Np+u) = \ln Np + u/Np - u^2/2(Np)^2 \quad \text{etc.}$$

which is legitimate if u/Np is small enough, we obtain, after some tiresome algebra,

$$\ln P = \frac{-u^2}{2Npq} + \ln\frac{1}{(2\pi Npq)^{1/2}} + \frac{u(p-q)}{2Npq} + \frac{u^2}{4}\left[\frac{1}{(Np)^2} + \frac{1}{(Nq)^2}\right]$$

where terms in u^2 only have been retained. For $Np, Nq \gg 1$, the last term can be neglected.

Finally, in this approximation, since $(p-q) < 1$ and is $\ll u$ nearly everywhere.

$$P = \frac{1}{(2\pi Npq)^{1/2}}\exp\left(\frac{-u^2}{2Npq}\right) \qquad 1.34$$

We shall discuss the physical significance of this distribution in later chapters: for the moment we consider it as a mathematical object. First, let us show that

$$\int_{-\infty}^{+\infty} P(u)\,du = 1$$

Put $\alpha = Npq$ and $w^2 = u^2/2\alpha$; then since the integral of

$$\surd(2\alpha)\int_{-\infty}^{+\infty}\exp(-w^2)\,dw = \surd(2\alpha)\pi$$

the integral

$$\int_{-\infty}^{+\infty} P(u)\,du = 1$$

This shows that the distribution is a proper probability function.

The evaluation of the integrals required to find $\langle u\rangle$ and $\langle u^2\rangle$ is dealt with in Appendix 1. The results are

$$\langle u\rangle = Np \qquad 1.35$$

$$\langle u^2\rangle = Npq \qquad 1.36$$

so the distribution has the same mean and variance as the binomial distribution.

*Stirling's full formula is $s! = \surd(2\pi s)\, s^s e^{-s} e^{\theta_s}$, with $\exp(\theta_s)$ the remainder term, defined by $|\theta_s| \leqslant 1/12s$. As stated, the remainder can be included to improve accuracy for small s; but for $s \geqslant 10$, the remainder is negligible.

In applications it is often useful to treat the *unit normal distribution*

$$P(x) = \frac{1}{\sqrt{(2\pi)}} \exp\left(-\frac{x^2}{2}\right)$$

which has zero mean and unit variance.

It is frequently required to find the probability that an observed value shall be less than a specified limit. This probability, for a continuous random variable u, is

$$\int_{-\infty}^{x} P(z)\,\mathrm{d}z$$

where x is the specified limit.

If this is written in the form

$$F(x) = \int_{-\infty}^{x} P(z)\,\mathrm{d}z$$

$F(x)$ is called the 'cumulative distribution function'. (This terminology is unfortunately not universal: statisticians sometimes call $F(x)$ the 'distribution function' and $P(z)$ the '(probability) density function.') The c.d.f. for the Gaussian distribution is

$$F(x) = \frac{1}{\sqrt{(2\pi)}} \int_{-\infty}^{x} \exp\left(-\frac{z^2}{2}\right) \mathrm{d}z$$

However, the *error function* $\operatorname{erf} t = \frac{2}{\sqrt{\pi}} \int_0^t \exp(-y^2)\,\mathrm{d}y$ is extensively tabulated.

Substituting $y^2 = z^2/2$

$$\operatorname{erf} t = \frac{2}{\sqrt{(2\pi)}} \int_0^{\sqrt{2}t} \exp\left(-\frac{z^2}{2}\right) \mathrm{d}z$$

or $\operatorname{erf} t = 2F(\sqrt{2}\,t) - 1$

and $F(\sqrt{2}\,t) = \frac{1}{2}\{\operatorname{erf}(t) + 1\}$

1.10 The central-limit theorem

We have shown in the foregoing examples that distributions which differ widely from one another for small numbers of events reduce to the same Gaussian distribution for very large numbers of events. It is natural to wonder whether this property is generally valid. The central-limit theorem of classical statistics provides the generalization required. The theorem is as follows.

If $x_1, x_2, x_3, \ldots, x_n$ are a sequence of independent random variables, the mean and variance can be calculated arithmetically. Let the mean of $x_n = \langle x_n \rangle$ the variance of $x_n = V_n$. Let $X = \Sigma x_n$. Then

$$Y = \frac{X - \sum_{i}^{s} \langle x_n \rangle}{\sqrt{\left(\sum_{i}^{s} V_n\right)}} \qquad 1.37$$

has a unit normal Gaussian distribution; that is, the probability distribution of x_n is given by

$$P(m, n) = \frac{1}{(2\pi\, V_n)^{1/2}} \exp\left[\frac{-(m - \langle x_n \rangle)^2}{2\, V_n}\right] \qquad 1.38$$

It is easy to show that this result is obtained using a two-term approximation to the characteristic function, but the theorem is much more general than this implies. The theorem is proved in many books on probability theory, for example Gnedenko.[3] The proofs are advanced and are mainly concerned with establishing the conditions under which the theorem is valid. A mathematically rigorous proof is given by Khinchin,[7] who establishes that, if moments up to the fifth are finite and are taken into the computation, equation 1.38 should be completed by an error term of order

$$O\left[\frac{1 + |m - \langle x_n \rangle|}{n^{3/2}}\right]$$

This is negligible for applications in statistical mechanics, because $n^{3/2}$ is a very large number indeed. The central-limit theorem forms the foundation stone for Khinchin's reformulation of statistical mechanics on the basis of modern probability theory.

2

Kinetic Theory and the Maxwell Distribution

The idea that a gas consists of an assembly of small mobile mass particles is a very old one and is due to the Greek philosophers. This idea was from time to time revived, but most scientists rejected it. In the middle of the nineteenth century, in spite of the general rejection of atomistic thinking, Clausius, Maxwell and Boltzmann succeeded in giving a kinetic theory of the motion of gas molecules which both had a reasonably solid mathematical foundation and gave many results which could be compared with experiment. When such comparisons were made it was found that the agreement between experiment and the theory was excellent in simple cases, although in others, especially when applied to gas mixtures, it was quite inadequate. Maxwell's work on the subject was published between 1859 and 1879 (therefore coming after his main work on electromagnetic theory) and by the 1880s kinetic theory was widely accepted, though less so in Germany than in other scientific communities. Boltzmann, working ten years later, very much improved the mathematical basis of the subject but, because of the resistance of the German/Dutch scientific establishment to kinetic ideas, his work was subjected to intense criticism. The criticisms are now known to be ill-founded and the critics are remembered only because of their part in this controversy, but the idea was propagated that Boltzmann's work was difficult and unreliable. This estimate has been overthrown only in the last twenty years by the rapid development of modern kinetic theory, which is the theory describing the transition of statistical assemblages of particles from non-equilibrium states to equilibrium.

This chapter gives an outline of Maxwell's formulation of kinetic theory, which is very easy to understand. The theory is then used to derive several results which are useful in themselves and which also demonstrate how the theory is used to derive numerical values for measurable quantities. Modern kinetic theory is introduced later in the book, and this chapter will help to clarify some of the problems.

2.1 The statistical nature of an ideal gas

From the viewpoint of statistical mechanics, an ideal gas is a gas whose particles interact with one another only during 'collisions'. The time spent in a collision is short by comparison with the time lapse between collisions. In the time between collisions, the gas particles move as free particles. Simple real gases approximate in their behaviour to ideal gases provided that the pressure is not too high. This criterion can be stated more precisely in the form that the particle density is so low that individual particles are relatively distant from all other particles, except for the short time spent in collisions.

The probability distribution function for the speed, component velocities, energy, etc. can be deduced for the ideal gas without making any assumptions about the force law which acts between particles when they collide. In fact, Maxwell himself carried out many derivations using the assumption that the force between particles varied as r^{-5}, where r is the separation between the two mass centres. This assumption eased the mathematics very much and in Maxwell's time it was thought that experimental evidence favoured a law of this form. This is now known to be incorrect and here we visualize, for the moment, the collisions as taking place between rigid, elastic spheres. Collisions also take place with the walls of the gas container, and it is assumed that the gas molecules obey the laws of elastic reflection.

We are now able to say something about the trajectory of an individual molecule. If we could choose a particle located at (r_0, t_0) and follow its motion, it would move in a straight line until a collision occurred, when the direction and velocity would be sharply changed. At the second collision, similar sudden changes of direction and velocity would take place. The succession of collisions with other particles goes on until a collision with a wall happens. The molecule is then reflected and starts a new progression of collisions with molecules until another wall collision occurs, and so on *ad infinitum*. Computer studies of trajectories using the Monte Carlo technique have been made and these give some insight into transient problems, for example the removal of a partition between gas and high vacuum.[1] However, the computer is incapable, in principle as well as in practice, of solving the dynamical problem set by enquiring about the trajectories.

It is important to understand why computers cannot be used, and we digress somewhat so as to dispose of this point. The number of molecules in a kilomole of gas,is, by definition, Avogadro's number, viz $6.022\,52 \times 10^{26}$; therefore, even a small volume of gas can contain 10^{20-22} molecules. Each molecular trajectory through a point has three initial velocity conditions so that three times as many calculations have to be started and followed up to the first collisions, and so on. Thus, at best, an enormous reduction factor has to be used so that the number of equations comes within the range of even the biggest computers. Even worse, though, is the fact that collisions of higher order occur in which *more* than two molecules are close enough to one another to interact simultaneously. Present-day dynamics cannot even solve the three-body collision problem without making simplifying assumptions; collisions between four and more bodies cannot be tackled at all. Lastly, even if one reduces the number of particles to figures which are unrealistically low and ignores the many-body collisions, it is still true that the trajectories themselves do not contain the kind of information physicists require. The situation is similar to that in fluid mechanics, where the Eulerian formulation is generally more useful than the Lagrangian one.

We are thus forced to seek new and less detailed ways of describing the behaviour of the gas. Here, a great simplification is achieved by restricting ourselves to a study of the state of thermal equilibrium. It is a matter of simple observation that the pressure, molecular number density, etc. of the gas depend not at all on the manner in which the system of gas-plus-container was set up. It does not matter at all if the gas was introduced fast or slowly, through one nozzle or two, or through some form of diffuser. The equilibrium state in which the gas in the container is characterized by a uniform pressure is achieved after a short time. Thus, for the moment, we ignore transient problems and concentrate on the equilibrium state.

In equilibrium, then, the molecules will be distributed over the volume of the vessel in a uniform manner, such that, if the volume is V and the total number of molecules is N, the density of molecules is N/V independently of the coordinates of the point where the measurement is made. Similarly, the distribution of molecules moving through a point in any direction must be isotropic, otherwise the particles would drift in a preferred direction and the density assumption would break down. However, it may be objected that many ways of injecting the gas *do* lead to a preferred direction. What factor destroys the ordered motion? The answer is, the collisions. Collisions with the walls reverse the directions of molecules which collide with incoming molecules and, after a lapse of sufficient time, the state of molecular chaos is reached. This remark emphasizes the somewhat paradoxical situation that, although the detailed nature of the collisions is not important, the fact that there *are* very many collisions is vital. It is also clear that measurements of the speed, or velocity component, made *after* the equilibrium state is reached must yield a *range* of velocities and not a single value of, for example, v_x. To verify this, consider that all the particles have a unique v_x; any collision will then yield particles with different values of v_x and the assumed state will be destroyed. Thus, not only are the particle velocities at any instant described by a (continuous) probability distribution function, but, supposing for the moment that we are able to label individual molecules, the members of the group of particles in Δx at v_x, say, are always changing because collisions scatter particles out of the group only to be replaced by equal numbers scattered into the group. This, by the way, is an example of what is called 'detailed balancing'.

2.2 The Maxwell distribution

Maxwell first calculated the law of distribution of velocities for an ideal gas and, according with what was said in the last section, he assumed that the gas was in equilibrium so that the state functions were all independent of time. Maxwell's original argument is given in a readily accessible form in reference 2. Here we follow the derivation of Levich[3], which is still both simple and elegant while at the same time keeping the role of collisions in view.

Let the velocity components of a molecule be u, v, and w; and dn_s the mean number density of molecules whose velocity components are in du at u, dv at v, dw at w. We have already said that the p.d.f must be isotropic in the velocities. Therefore the velocity distribution function $n(\mathbf{v})$ can only be a function of the argument

$$s = |\mathbf{v}| = (u^2 + v^2 + w^2)^{\frac{1}{2}} \tag{2.1}$$

$$\therefore \; dn_s \equiv dn(\mathbf{v})$$

and we can write

$$\begin{aligned} dn_s &= n(\mathbf{v})\, du\, dv\, dw \\ &= n(s)\, du\, dv\, dw \end{aligned} \tag{2.2}$$

Consider now collisions between two particles with velocities $\mathbf{v}_1$ and $\mathbf{v}_2$. They collide and the velocities after collision are $\mathbf{v}_3$ and $\mathbf{v}_4$. Since energy is conserved, we immediately have

$$s_1^2 + s_2^2 = s_3^2 + s_4^2 \qquad 2.3$$

Moreover, the number of molecules with velocities $\mathbf{v}_1$ is proportional to $n(s_1)$ and the number with $\mathbf{v}_2$ is proportional to $n(s_2)$. Thus, the number of collisions of the $\mathbf{v}_1 \rightarrow \mathbf{v}_2$ type, per unit volume, is proportional to $n(s_1) \,.\, n(s_2)$.

But, since the number of particles in the $\mathbf{v}_1$, $\mathbf{v}_2$ class is conserved, by detailed balance, we must conclude that the number density of $\mathbf{v}_3$, $\mathbf{v}_4$ collisions is such that

$$n(s_1) \,.\, n(s_2) = n(s_3) \,.\, n(s_4) \qquad 2.4$$

It very much simplifies matters if we rewrite equation 2.4 in the form

$$n(s_1^2) \,.\, n(s_2^2) = n(s_3^2) \,.\, n(s_4^2) \qquad 2.5$$

which is obviously allowed since the distribution functions depend only on the values of s. We can now use equation 2.3 to transform equation 2.5 to

$$n(s_1^2) \,.\, n(s_2^2) = n(s_3^2) \,.\, n(s_1^2 + s_2^2 - s_3^2)$$

or, taking logs,

$$\ln n(s_1^2) + \ln n(s_2^2) = \ln n(s_3^2) + \ln n(s_1^2 + s_2^2 - s_3^2) \qquad 2.6$$

We can now differentiate with respect to s_1^2 and s_2^2, remembering that s_1, s_2, s_3 are completely arbitrary since s_4 has been eliminated by equation 2.3. The result is

$$\frac{1}{n(s_1^2)} \cdot \frac{dn(s_1^2)}{d(s_1^2)} = \frac{1}{n(s_1^2 + s_2^2 - s_3^2)} \cdot \frac{dn(s_1^2 + s_2^2 - s_3^2)}{d(s_1^2 + s_2^2 - s_3^2)}$$

$$\frac{1}{n(s_2^2)} \cdot \frac{dn(s_2^2)}{d(s_2^2)} = \frac{1}{n(s_1^2 + s_1^2 - s_3^2)} \cdot \frac{dn(s_1^2 + s_2^2 - s_3^2)}{d(s_1^2 + s_2^2 - s_3^2)}$$

or

$$\frac{1}{n(s_1^2)} \cdot \frac{dn(s_1^2)}{d(s_1^2)} = \frac{1}{n(s_2^2)} \cdot \frac{dn(s_2^2)}{d(s_2^2)} \qquad 2.7$$

Equation 2.7 can be true only if the right-hand and left-hand sides both equal a common constant, which we call $-\alpha$. Integrating, we get

$$n(s_1^2) = A \exp(-\alpha s_1^2) \qquad 2.8$$

A is a normalization constant, which must be evaluated by the condition that the total number density of molecules present is $N/V = n$. The normalization is most easily carried out by rewriting equation 2.2 in spherical polar coordinates, s, θ, ψ. Then

$$dn_s = n(s)\, s^2\, ds \,.\, \sin\theta \, d\theta \, d\psi \qquad 2.9$$

The limits of ψ are 0, 2π and those of θ are $-\pi \rightarrow +\pi$, so we can carry out two integrations to get

$$n = 4\pi \int n(s)\, s^2\, ds \tag{2.10}$$

The lower limit of s is clearly zero. The upper limit is indefinite, by relativity it is presumably $< c$, the velocity of light. However, the exponential fall off of the integrand is so fast that we can take the integral between 0 and ∞ without error. Then

$$n = 4\pi A \int_0^\infty s^2 \exp(-\alpha s^2)\, ds \tag{2.11}$$

From equation 2.11 it is necessary that $\alpha > 0$, otherwise the integral does not converge. Integrating equation 2.11 we get

$$A = n(\alpha/\pi)^{\frac{3}{2}} \tag{2.12}$$

Finally, using equation 2.12 we obtain

$$n(s) = n(\alpha/\pi)^{\frac{3}{2}} \exp(-\alpha s^2) \tag{2.13}$$

and

$$dn(s) = n\,.\,4\pi(\alpha/\pi)^{\frac{3}{2}} s^2 \exp(-\alpha s^2)\; ds \tag{2.14}$$

by, for example, differentiating equation 2.10. Equation 2.14 gives the number-density distribution function for molecules in the speed range ds at s. By transformation back into rectangular coordinates, we get

$$dn(s)\; n\,.\,(\alpha/\pi)^{\frac{3}{2}} \exp[-x(u^2 + v^2 + w^2)]\;\; du\, dv\, dw \tag{2.13}$$

which can be written

$$\begin{aligned} dn(s) &= dn(u)\,.\,dn(v)\, dn(w) \\ &= n(\alpha/\pi)^{\frac{1}{2}} \exp(-\alpha u^2)\, du\,.\,(\alpha/\pi)^{\frac{1}{2}} \exp(-\alpha v^2)\, dv\,.\,(\alpha/\pi)^{\frac{1}{2}} \exp(-\alpha w^2)\, dw \end{aligned}$$

in which the velocity distribution function is expressed as a *product* of the three velocity-component distributions. Frequently we wish to know, say, the u distribution for arbitrary v and w. Since

$$\int_{-\infty}^{+\infty} \exp(-\alpha\zeta^2)\, d\zeta = \sqrt{(\pi/\alpha)}$$

the result is

$$\mathrm{d}n(u) = n(\alpha/\pi)^{\frac{1}{2}} \exp(-\alpha u^2)\, \mathrm{d}u \qquad 2.15$$

Here, the limits are assigned remembering that v and w can be directed in either sense.

It is worth remarking that these distribution formulae are *not* probability density distributions because they are normalized so as to provide for an, assumed known, molecular number density instead of being normalized to unity. Once again, it should be remembered that it is necessary to keep the normalization constants distinguished from one another.

In the formulae derived so far, we still have the unknown constant α. How do we assign a value to α? This is done by using Maxwell's distribution to calculate one of the physical properties of the system (gas + container) which can be compared either with experiment or with the results of independent theory. The most usual calculation is to use the equation of state

$$pV = NkT$$

$$\text{or} \quad p = nkT \qquad 2.16$$

which relates the pressure with the number density, Boltzmann's constant and the absolute temperature. This gives the result

$$\alpha = m/2kT \qquad 2.17$$

as we shall demonstrate in the next section.

2.3 The kinetic calculation of the pressure

The basis of the calculation is very simple. According to the assumptions made about molecule–wall collisions, the velocity of an incident molecule (u, v, w) is changed to $(-u, v, w)$ on making a wall collision. The change in momentum, $2mu$, is transferred from the molecule to the wall. The resultant force per unit area on summing over all wall collisions is equated to the pressure.

We have already calculated the number density of particles $\mathrm{d}n(u)$ in $\mathrm{d}u$ at u. To hit, during unit time, a wall which is normal to u, a particle with velocity u_1 must be located in a parallelepiped of volume $u_1 \times 1$, where the cross-sectional area is unity. The number of molecules in this parallelepiped is

$$\Delta n = n(\alpha/\pi)^{\frac{1}{2}}\, u_1 \exp(-\alpha u_1^2)\, \mathrm{d}u_1$$

The momentum transferred is $\Delta n\, .\, 2mu$. Thus, this group of molecules transfers

$$\Delta M = n\, .\, 2m(\alpha/\pi)^{\frac{1}{2}}\, u_1^2 \exp(-\alpha u_1^2)\, \mathrm{d}u_1$$

to the wall, per unit area, per second. The total transfer is obtained by integrating from $u_1 = 0$ to $u_1 = \infty$, since only one sense of motion is utilized. This transfer is equal to the force, which in this case is called the pressure p. Therefore

$$p = n \,.\, 2m \,.\, (\alpha/\pi)^{\frac{1}{2}} \int_0^\infty u_1^2 \exp(-\alpha u_1^2)\, du_1$$

The integral equals $\pi^{\frac{1}{2}}/4\alpha^{\frac{3}{2}}$, so we find

$$p = nm/2\alpha \qquad 2.18$$

Comparing equation 2.18 with equation 2.16, we find $\alpha = m/2kT$ as stated in equation 2.17.

We have now obtained expressions for the speed and velocity distributions which depend only on measurable parameters and universal constants. In fact, we need only know the pressure and the absolute temperature and have a value of k for a complete numerical specification of the distribution function, since, from equation 2.16,

$$n = p/kT \qquad 2.19$$

At this stage it is worthwhile to consider exactly what has and has not been proved by the above development. The proof shows that, once the gas has reached a state of equilibrium described by the Maxwellian distribution, detailed balancing ensures that it will remain in that state. It does *not* show that the gas will automatically reach the Maxwellian distribution or, in other words, it has not been shown that the Maxwellian distribution is unique. The Maxwellian actually turns out to be unique, but the proof relies on fairly advanced considerations of probability theory applied to Gaussian distributions. These developments were not available to Maxwell or to Boltzmann, and their lack was at the root of the reluctance to accept Boltzmann's developments of Maxwell's work. We shall return to this topic when we introduce modern kinetic theory: for the moment, let us accept the Maxwellian and pursue the question of calculating useful quantities by its use.

2.4 Applications of the Maxwellian distribution

For ease of reference, let us recapitulate the results so far, replacing α by $(m/2kT)$. The number of molecules with speeds in ds at s is

$$dn_s = 4\pi n(m/2\pi kT)^{\frac{3}{2}}\, s^2 \exp(-ms^2/2kT)\, ds \qquad 2.20$$

This can be written in terms of the component velocities as

$$dn_s = n\left(\frac{m}{2\pi kT}\right)^{\frac{3}{2}} \exp\left[-\frac{m}{2kT}(u^2 + v^2 + w^2)\right] du\, dv\, dw \qquad 2.21$$

The number density of particles with u velocity in du at u is

$$dn_{(u)} = n(m/2\pi kT)^{\frac{1}{2}} \exp(-mu^2/2kT)\ du \qquad 2.22$$

A new result which is often useful is the distribution in terms of kinetic energy. Since $E = \frac{1}{2}ms^2$, $\quad dE = ms\ ds$

and
$$dn_{(E)} = \frac{2n}{\pi^{\frac{1}{2}}}\left(\frac{1}{kT}\right)^{\frac{3}{2}} \exp\left(-\frac{E}{kT}\right) E^{\frac{1}{2}}\ dE \qquad 2.23$$

The corresponding *probability* distribution functions are clearly obtained by dividing any of these four equations by n.

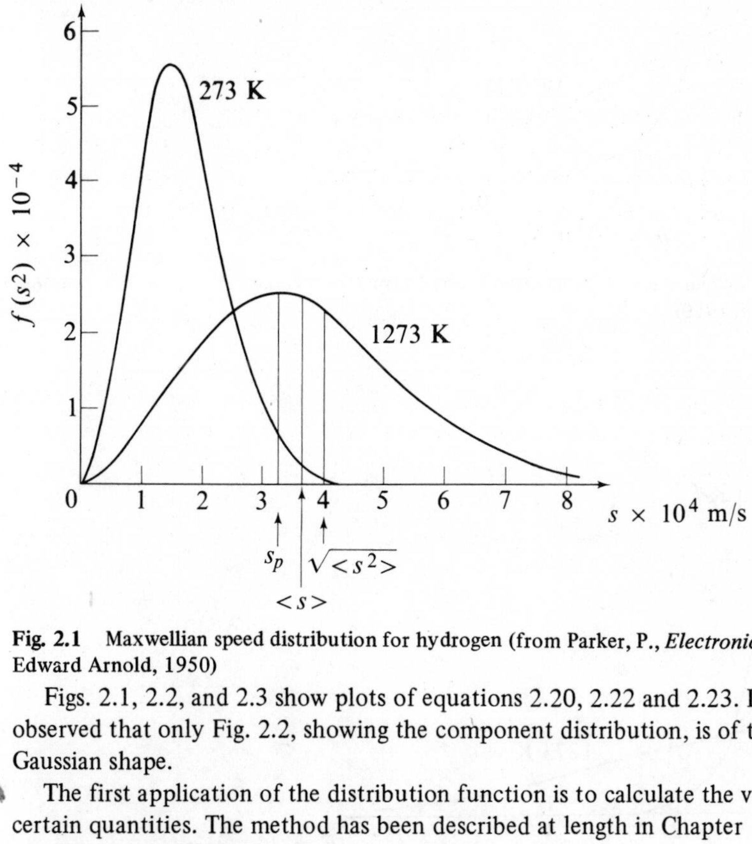

Fig. 2.1 Maxwellian speed distribution for hydrogen (from Parker, P., *Electronics,* London, Edward Arnold, 1950)

Figs. 2.1, 2.2, and 2.3 show plots of equations 2.20, 2.22 and 2.23. It will be observed that only Fig. 2.2, showing the component distribution, is of the typical Gaussian shape.

The first application of the distribution function is to calculate the values of certain quantities. The method has been described at length in Chapter 1; as an example, we shall calculate the mean speed $\langle s\rangle$.

By definition,

$$\langle s\rangle = \frac{1}{n}\int_0^\infty dn_s\ ds$$

$$= 4\pi\left(\frac{m}{2\pi kT}\right)^{\frac{3}{2}} \int_0^\infty s^3 \exp\left(-\frac{ms^2}{2kT}\right) ds$$

The integral can easily be evaluated (Appendix 1) and is $2(kT/m)^2$. Therefore

$$\langle s\rangle = 2\sqrt{(2/\pi)}\ .\ (kT/m)^{\frac{1}{2}} \qquad 2.24$$

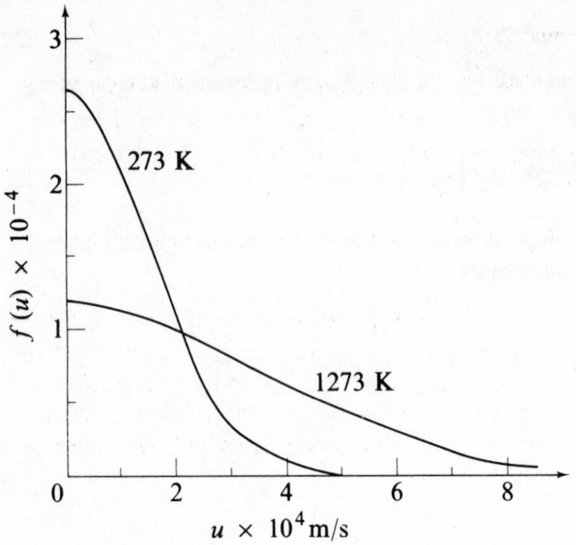

Fig. 2.2 Maxwellian velocity distribution for hydrogen (from Parker, P., *Electronics*, London, Edward Arnold, 1950)

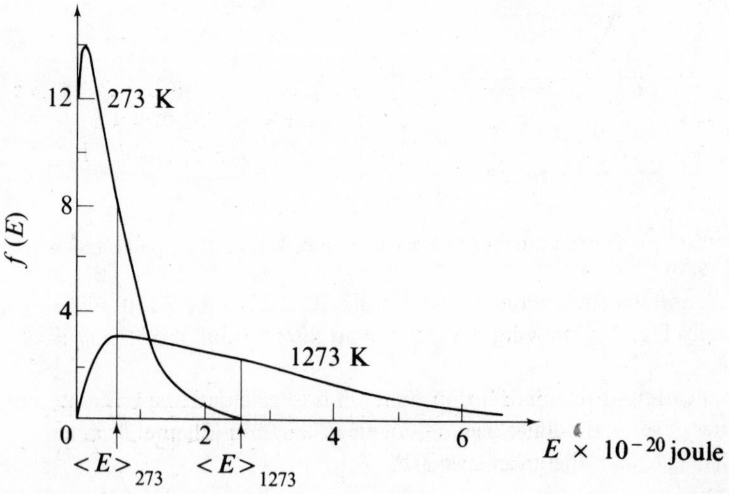

Fig. 2.3 Maxwellian distribution of kinetic energy for hydrogen (from Parker, P., *Electronics*, London, Edward Arnold, 1950)

Exercise

Prove the following results:

1. $\langle u \rangle = 0$

(this is obvious by inspection of Fig. 2.2)

2. $\langle s^2 \rangle = 3/2 \,.\, (2kT/m)^{\frac{1}{2}}$
3. $\langle E \rangle = 3kT/2$
4. The most probable speed is $(2kT/m)^{\frac{1}{2}}$

Result 3 is very important: it is the basis of the classical equipartion of energy theorem. Since our hard-sphere molecules have only three degrees of freedom, the result is usually expressed by saying that, on average, an energy $\frac{1}{2}kT$ is associated with each degree of freedom of a classical particle.

2.5 Mean free path and collision frequency

Before using the Maxwellian distribution to make elementary calculations on the transport quantities, (viscosity, thermal and electric conductivity, etc.) in a slightly non-uniform gas, we must consider collision processes in a little more detail. First, we define the total collision cross-section, symbol σ. For our hard, spherical molecules, Fig. 2.4 shows the situation when a molecule A of radius a collides with a

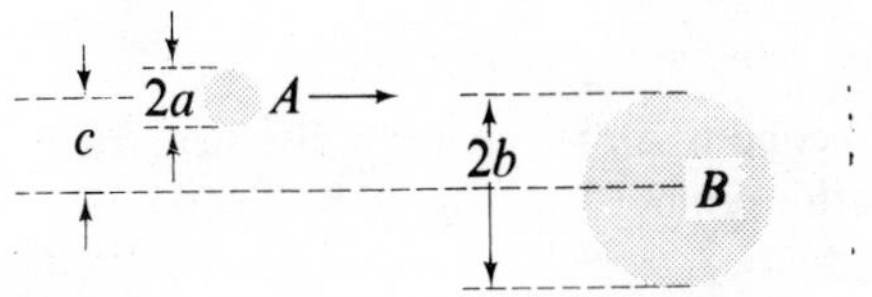

Fig. 2.4 Notation for a hard sphere collision.

molecule of a different species, B radius b. If the displacement c between the respective centres is $> (a + b)$, there is no collision. On the other hand, if $c < (a + b)$, a collision occurs. Thus, a collision happens only if the centre of B lies inside the volume whose cross-sectional area is

$$\sigma = \pi(a + b)^2 \qquad 2.25(a)$$

or, for identical molecules,

$$\sigma = 4\pi a^2 = \pi d^2 \qquad 2.25(b)$$

Next, we wish to evaluate the mean relative speed of pairs of particles in a collision. This calculation needs a little care. Let the colliding particles have velocities $\mathbf{v}_1$ and $\mathbf{v}_2$. Then, the relative velocity is

$$\mathbf{V} = \mathbf{v}_1 - \mathbf{v}_2 = \dot{\mathbf{r}}$$

We must remember that, to analyse the collision dynamics simply*, we have to work in a moving-centre-of-mass system and that the relative motion occurs in the Lagrangian in the form $\frac{1}{2}\mu(\dot{\mathbf{r}})^2$, where $\mu = m_1 m_2/(m_1 + m_2)$ and $\dot{\mathbf{r}}$ is the relative velocity. μ is called the reduced mass and is $m/2$ for identical particles. Thus, the

*Collision dynamics will be discussed in detail when we come to charged particles.

distribution for the relative speed is Equation 2.20 with $m/2$ replacing m and

$$\langle s_{\text{rel.}}\rangle = 4\pi\left(\frac{m}{4\pi kT}\right)^{\frac{3}{2}} \int_0^\infty s_{\text{rel.}}^3 \exp\left(-\frac{ms_{\text{rel.}}^2}{4kT}\right) ds_{\text{rel.}}$$

Carrying out the integration,

$$\langle s_{\text{rel.}}\rangle = \frac{4}{\sqrt{\pi}}\left(\frac{kT}{m}\right)^{\frac{1}{2}} = \sqrt{2}\,\langle s\rangle \qquad 2.26$$

The total number of collisions made per second by the test molecule travelling through a background of stationary, identical molecules, is

$$n\sigma\langle s_{\text{rel.}}\rangle = 4n\sigma(kT/\pi m)^{\frac{1}{2}}$$

(It is here assumed, correctly in the simple model we are discussing, that σ does not depend on s.) The total number of collisions per second is called the 'collision frequency', symbol ν. So we have

$$\nu = 4n\sigma(kT/\pi m)^{\frac{1}{2}} \qquad 2.27$$

The length of the path travelled in one second is $\langle s\rangle$, so the mean distance between collisions, called the 'mean free path', symbol λ, is seen to be given by

$$\lambda = \langle s\rangle/\nu = 1/\sqrt{2}(n\sigma) \qquad 2.28$$

The mean free time is sometimes used. This is merely

$$\frac{1}{\nu} = \frac{1}{4n\sigma}\left(\frac{\pi m}{kT}\right)^{\frac{1}{2}} \qquad 2.29$$

It is perhaps more useful to express λ in terms of the pressure, instead of n. Using Equation 2.19,

$$\lambda = kT/\sqrt{2}(\sigma p) \qquad 2.30$$

For nearly all monatomic gases at room temperature and $p = 0.1$ *Pa* (the sort of value reached by a mechanical vacuum pump), λ does not differ from 60 mm by more than a factor of 2. A knowledge of this magnitude is helpful in deciding whether binary collisions between molecules are dominant or whether the pressure is low enough for molecule–wall collisions to play a major role. For example, in a long cylindrical vessel 50 mm diameter at 10^{-5} torr, molecular collisions with the walls are much more frequent than binary collisions. Such considerations are important in flow problems.

To conclude this section we shall calculate the probability that a molecule will travel a length l *without* making a collision. Let $p(l)$ equal this probability. But the probability of a collision being made in an infinitessimal distance dl can be written as $a.dl$, where a is a proportionality constant, which we shall determine. The probability of *no* collision in dl is then, by definition,

$$p(dl) = 1 - a.dl$$

Thus, we can write, since the events are independent,

$$p(l + \mathrm{d}l) = p(l) \, . \, p(\mathrm{d}l) = p(l) \, (1 - a \, . \, \mathrm{d}l)$$

for the probability of moving through $l + \mathrm{d}l$ without collision

$$\mathrm{d}p(l)/\mathrm{d}l = -a \, . \, p(l)$$

or $$p(l) = A \exp(-al)$$

But $p(0)$ is obviously 1.0, so $A = 1$.

$$\therefore \; p(l) = \exp(-al) \qquad 2.31$$

If we define $\lambda = \langle l \rangle$, we obtain

$$\lambda = \frac{\int_0^\infty l \, p(l) \, \mathrm{d}l}{\int_0^\infty p(l) \, \mathrm{d}l} = \frac{\int_0^\infty l \exp(-al) \, \mathrm{d}l}{\int_0^\infty \exp(-al) \, \mathrm{d}l}$$

$$= 1/a$$

Therefore, the final result is

$$p(l) = \exp(-l/\lambda) \qquad 2.32$$

The probability of a collision in $\mathrm{d}l$ is

$$\mathrm{d}p = \frac{1}{\lambda} \exp\left(-\frac{l}{\lambda}\right) \mathrm{d}l \qquad 2.33$$

It is sometimes incorrectly stated that Equation 3.32 applies only from the point at which the last collision took place. In fact, it applies to the distance from any initial point at which the molecule was certainly located, i.e. it is a conditional probability contingent on locating the molecule at $l = 0$.

2.6 Introductory transport theory

We have now collected enough theoretical apparatus to discuss, in an elementary way, how small departures from the Maxwellian equilibrium give rise to the flow of certain material properties of the gas. It is perhaps surprising that what is effectively a perturbation theory on the distributions should give many results which are in good agreement with experiment, but such is the case. However, we are at present mainly interested in showing how the methods work in a context which minimizes purely mathematical difficulties.

The basic procedure is to assume, for small departures from equilibrium, that the various distribution functions are still Maxwellian.

Example 1. Flow through a small leak

Consider a small hole, area $\Delta\Sigma$, in a thin diaphragm separating two gas enclosures,

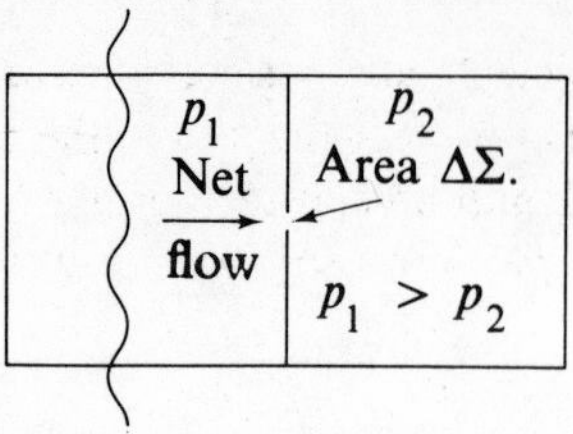

Fig. 2.5 Gas efflux through an aperture.

initially at pressures p_1 and p_2 (Fig. 2.5). If the mean free paths λ_1 and λ_2 at the respective pressures are both long compared with the hole dimensions, all molecules reaching the hole can be assumed to penetrate it.

In section 2.3 we showed that the number of molecules per unit time per unit area is obtained by integrating $\Delta n_1 = n_1(\alpha/\pi)^{\frac{1}{2}} \exp(-\alpha u_1{}^2)\, u_1\, du_1$ from zero to infinity. Call the result N_1, and assume $p_1 > p_2$; then the mass flow per second from $p_1 \to p_2$ is $MN_1\Delta\Sigma$, where M = molecular mass. Similarly, there is a counter flow $MN_2\Delta\Sigma$ from $p_2 \to p_1$. Thus, the net mass flow from $p_1 \to p_2$ is $(N_1 - N_2)M\Delta\Sigma$. Then,

$$N_1 = n_1\left(\frac{\alpha}{\pi}\right)^{\frac{1}{2}} \int_0^\infty u_1 \exp(-\alpha u_1{}^2)\, du_1$$

$$= \frac{n_1}{2}\left(\frac{1}{\alpha\pi}\right)^{\frac{1}{2}} \qquad 2.31$$

Similarly,

$$N_2 = \frac{n_2}{2}\left(\frac{1}{\alpha\pi}\right)^{\frac{1}{2}}$$

But, $n_1 = 2\alpha p_1/M, \quad n_2 = 2\alpha p_2/M, \quad \alpha = M/kT,$ therefore

$$\text{mass flow per second} = (p_1 - p_2)\, \Delta\Sigma (M/\pi kT)^{\frac{1}{2}} \qquad 2.32$$

Vacuum technologists express this result in various units, but the important point is that the mass flow rate is directly proportional to the pressure difference and to the area and, through M, depends on the gas. Clearly, expressions of this type are useless if the area is too large, because the pressure would rapidly equalize and the mass flow rate would drop to zero. However, if p_2 is maintained at a nearly constant value with a vacuum pump, equation 3.32 gives a useful estimate of the leak rate and the time it takes to pump p_1 down to p_2. Before leaving this calculation, we should notice that equation 3.31 for N can be written differently, in terms of the mean velocity. In fact,

$$N_1 = n_1\langle s\rangle/4\sqrt{2} \text{ per unit area, per second} \qquad 2.33$$

(To avoid possible confusion, it must be noted that very elementary treatments of transport theory replace equation 2.33 by $N_1 = n_1\langle s\rangle/6$, which is only about 6% different from equation 3.33. This result comes from reasoning that roughly $n_1/3$

molecules have mean speed $\langle s\rangle$ along each of the three coordinate vectors; therefore, $n_1/6$ move to the right, i.e. in the z direction, and another $n_1/6$ to the left, in the $-z$ direction. Thus, $N_1 \approx n_1\langle s\rangle/6$ = number passing through unit area of the normal plane per second. Let us, from now on, put $\gamma = 1/6$ or $1/4\sqrt{2}$ to carry *both* results along.)

Example 2. Thermal conductivity of a gas

The thermal conduction coefficient is the constant κ in

$$Q_z = -\kappa \frac{\partial T}{\partial z} \qquad 2.34$$

where Q = heat flux in z direction. The minus sign is consistent with heat flow from the higher to the lower temperature region.

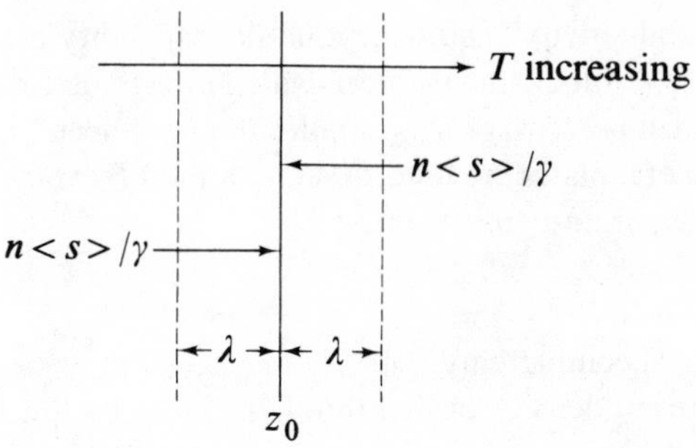

Fig. 2.6 Notation for heat transfer in a slightly non-uniform system.

The notation for the problem is shown in Fig. 2.6; z_0 represents a plane perpendicular to the direction of heat flow. The temperature is supposed to vary slowly, so that the pressure is constant over the region shown. The molecules are assumed to have taken on the characteristic energy of their position at the last collision before crossing the plane z_0. Thus, the temperature is assumed to vary over distances of the order of two mean free paths. The energy of the molecules then becomes a function of z. Let $\langle E_{(z)}\rangle$ denote the mean energy of molecules at z. The molecules passing through z from right to left have greater energies than those moving from left to right. Then

$$Q_z = \frac{n\langle s\rangle}{\gamma}\,[\langle E_{(z-\lambda)}\rangle - \langle E_{z+\lambda)}\rangle]$$

$$= \frac{n\langle s\rangle}{\lambda}\left[\langle E_{(z)}\rangle - \frac{\lambda\partial E}{\partial z} - \left(\langle E_{(z)}\rangle + \frac{\lambda\partial E}{\partial z}\right)\right]$$

$$= -\frac{n\langle s\rangle}{\gamma}\,.\,2\lambda\frac{\partial E}{\partial z}$$

$$= -\frac{2n\langle s\rangle\lambda}{\gamma}\cdot\frac{\partial E}{\partial T}\cdot\frac{\partial T}{\partial z} \qquad 2.35$$

Comparison with equation 2.34 gives

$$\kappa = \frac{2n\langle s\rangle\lambda}{\gamma} \cdot \frac{\partial E}{\partial T} \qquad 2.36$$

But $\partial E/\partial T = c$ = specific heat per molecule

$$\therefore \quad \kappa = \frac{2n\langle s\rangle\lambda c}{\gamma}$$

$$= \frac{2\langle s\rangle c}{\gamma\sqrt{2}\sigma} \qquad 2.37$$

on substituting for n and λ. Thus, κ is independent of the pressure. It varies as $T^{\frac{1}{2}}$ from $\langle s\rangle$, but experimentally temperature variations in c and σ cause faster variation.

Example 3. Electrical conductivity

Strictly, the calculation of electrical conductivity is quite beyond the capability of the theory developed so far. However, a result can be deduced which is both useful and often quoted. We assume that a small percentage of gas molecules have been ionized, for example by X-rays. Two metal plates are used to set up a field E_z volts/metre in the z direction. Then, working per unit area, we have

$$j_z = \sigma_e \,.\, E_z \qquad 2.38$$

where j_z = current density, σ_e = electrical conductivity.

Once more we can write, for the current density carried through a plane by the charged particles of density n_e,

$$j_z = n_e \,.\, e \,.\, \langle v_z\rangle \qquad 2.39$$

Here $\langle v_z\rangle$ is quite different from the mean molecular velocity $\langle s\rangle$ and is established by the following approximate reasoning. Consider that a charged molecule has collided with another molecule and has had its initial velocity fixed at v_0. The motion is then given by

$$M\dot{v}_z = eEz$$

$$\text{or} \quad v_z = \frac{eEz}{M} \,.\, t + v_0 \qquad 2.40$$

We now take an ensemble average, on which $\langle v_0\rangle = 0$ because v_0 is isotropically distributed. Then,

$$\langle v_z\rangle = \frac{eEz}{M} \,.\, \langle t\rangle \qquad 2.41$$

Here $\langle t\rangle$ is the mean time between collisions, which is usually given the symbol τ.

$$\text{Then} \quad \langle v_z\rangle = \frac{eEz}{M} \,.\, \tau$$

$$\text{and} \qquad \sigma_e = \frac{n_e \,.\, e^2 \,.\, \tau}{M} \qquad 2.42$$

The usefulness of this is limited by uncertainty about the value of τ. Collisions in this case can be either ion–molecule or ion–ion collisions. Under the assumptions, ion–molecule collisions are much more frequent than others. However, the form of equation 2.42 is interesting in many circumstances; for example, it shows why proportional counters work. Moreover, expressions exactly similar to equation 2.42 apply to semiconductors if the effective mass is used for M and is appropriately defined. We shall establish this result much more rigorously later in the book.

Example 4. Diffusion coefficient

Later in this book we shall find that diffusion plays an important role in modern non-equilibrium theory. We therefore introduce the concept in a very simple form here. Notice that, up to now, the non-equilibrium quantities we have discussed are independent of time. We assumed that by some external means or other we have ensured a departure from strict equilibrium and the calculations really reduce to calculations of the difference between two equilibrium states.

The phenomenon of diffusion introduces time-dependence, because the diffusion equation is of the one-dimensional form

$$\frac{\partial n}{\partial t} = D\frac{\partial^2 n}{\partial z^2} \qquad 2.43$$

where D = diffusion coefficient.

In establishing equation 2.43 it is assumed that some of the molecules in the gas are labelled with respect to the others; for example, a small quantity of an isotope could be released at a fixed point in the background gas. How do these labelled molecules drift out to achieve their equilibrium (uniform) distribution? Equation 2.43 describes the one-dimensional form of the process, which works as follows. When the labelled particles are released, they make collisions with the background gas as well as with one another; thus, they rapidly acquire the mean velocity characteristic of the whole system. They then move away from the point of release, pursuing a completely random trajectory until a new equilibrium is established. But this, after all, is just what establishes the Maxwellian distribution of the background molecules; thus, the labelled molecules behave exactly like the background, except that they can be kept track of.

Let J_z = mean number density of labelled particles crossing a z plane per unit time

Then $$\frac{\partial J_z}{\partial z} + \frac{\partial n_L}{\partial t} = 0 \qquad 2.44$$

and $$J_z = -D\frac{\partial n_L}{\partial z}$$ from equation 2.43.

Here n_L is the density of *labelled* particles. The situation is very similar to that depicted in Fig. 2.6, if we now remember that n_L is a function of z, instead of the energy in the earlier calculation. Then,

$$J_z = \frac{\langle s \rangle}{\gamma}\,[n_1(z-\lambda) - n_1(z+\lambda)]$$

$$= -\frac{2\langle s \rangle}{\gamma}\,.\,\lambda\frac{\partial n_1}{\partial z}$$

Therefore $D = 2\langle s \rangle \lambda / \gamma$ 2.45

Diffusion phenomena are at the heart of irreversible statistical mechanics, since the form of equation 2.43 shows that the solution must depend on the sign of t. For the first time we encounter a phenomenon in which it matters whether t increases or decreases. We shall later find that diffusion is closely connected with frictional phenomena and energy loss in the gas.

At this point, we have said enough about simple transport theory. Transport theory applied to non-ideal gases, to gas mixtures and other more complicated systems is exhaustively discussed by Chapman and Cowling[2], whose work is recommended to specialists in gas dynamics etc. Kinetic theory applied to vacuum technology is discussed in detail by Swift[4].

3

Statistical Mechanics by the Method of Gibbs

We now depart from the historical order of development, postponing any discussion of the contribution of Boltzmann until later, and move on to the work of J. W. Gibbs at the turn of the century. Gibbs' outstanding contribution was to describe methods of statistical treatment which apply to systems which are much more complicated than gas molecules and, in particular, which can be of macroscopic dimensions, with all that that implies in terms of internal structure, large numbers of degrees of freedom, etc.

Gibbs started with ensemble ideas, that is he imagined experiments conducted on an ensemble of very large numbers of identical systems. Probabilistic theory is used to predict the outcome of these experiments, particular attention being paid to the manner in which the total available energy can be apportioned to the macroscopic systems. As an example, Gibbs' work would cover an ensemble experiment carried out on a very large number of rigid rod pendula. The pendulum is, of course, a very well understood example of a simple harmonic oscillator, but the point to understand here is that even the smallest macroscopic pendulum must consist of extremely large numbers of microscopic particles, all with internal energy structures. Gibbs' technique allows statistical calculations about the pendula to be set up, without requiring any information about the detail of the microscopic structure.

Another important point is that it is quite simple to introduce quantum ideas into Gibbsian statistical mechanics. In fact certain simplifications result, if we do this, so in this book we shall find it better to avoid an artificial distinction between classical and quantum mechanics.

3.1 Representative point, phase space, velocity or momentum space, and configuration space

The reader is presumed to have some degree of acquaintanceship with classical mechanics developed by Hamilton's method – references 1 to 3 provide suitable treatments at increasingly advanced level.

The Hamilton equations express the relationship between the *Hamiltonian function* (H) and the coordinates (symbol q_k) and momenta (symbol p_k) of the body concerned. H may be thought of as the total energy of the system. It is a function of time, q's and p's. Hamilton's form of the equations of motion is, for f degrees of freedom,

$$\partial H/\partial p_k = \dot{q}_k \quad \text{for all } k \leqslant f \tag{3.1}$$

$$\partial H/\partial q_k = -\dot{p}_k \quad \text{for all } k \leqslant f \tag{3.2}$$

and thus constitutes a set of $2f$ *first*-order partial differential equations. For a system with N particles there are $3N$ degrees of freedom so that there are $f = 3N$ co-ordinates and $f = 3N$ momenta. Given a sufficient number of initial conditions, the $6N$ equations can be solved to give the way in which q's and p's evolve in time, that is the generalized trajectory of the system.

Clearly, it is impossible to represent a complicated system by a graphical trajectory, but as an aid to visualization, the idea of representative points in phase space is introduced. For a system of $3N$ degrees of freedom, phase space is a $6N$-dimensional hyperspace in which the axes are the p's and q's. Since the highest value of N which can readily be sketched is $n = 1$, giving a p_1, q_1 phase space, the concept is usually illustrated by an easy but non-trivial particular system, a single mass particle executing simple harmonic motion along the q_1 axis.

The reader will easily verify for himself that, for mass m and restoring force $-\kappa q_1$,

$$(p/m\omega A)^2 + (q/A)^2 = 1 \qquad 3.3$$

$$p^2/2m + \pi q^2/2 = E \qquad 3.4$$

where A = amplitude of the oscillation, $\omega = \sqrt{(\kappa/m)}$ = the angular frequency of the oscillation, and E is the total energy (the instantaneous sum of the ke and pe of the oscillator). Equation 3.4 is the equation of an ellipse with p semi-axis length $p = \sqrt{(2m)}E$ and q semi-axis $\sqrt{(2E/\kappa)}$.

Now, the total energy of the oscillator is *not* exactly known. There are two sources of indefiniteness. Firstly, there is the quantum-mechanical uncertainty principle which asserts that no measurements made on p and q simultaneously will yield a certain value for E; more strictly put, a value of E which is certain corresponds with small ranges of p and q. Secondly, and more important in many applications of the theory, it is wholly unrealistic to consider that the oscillator can be completely isolated from the rest of the world. In fact, it is (however weakly) coupled to the rest of the world, and a *small* interaction energy ΔE_i is involved. In other words, real oscillators are *damped* and can be maintained in a state of uniform oscillation only by the continuous supply of the interaction energy. However, if we inspect the oscillator for only a short time, the damping will not be observed. Instead, if we look at an ensemble of identical elementary oscillations, we shall find slightly different energies in the range $E + \Delta E_i$.

Fig. 3.1 shows the phase space for this problem, with the shaded area indicating

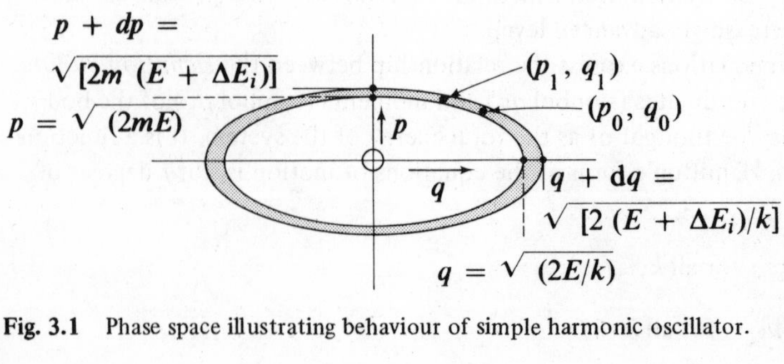

Fig. 3.1 Phase space illustrating behaviour of simple harmonic oscillator.

the region of phase space which is accessible to the simple harmonic oscillator with mean energy E. In other words, if at time t_0 we measure p and q for our oscillator, obtaining p_0 and q_0, we can represent this state of the oscillator by the point (p_0, q_0) in the shaded area. This is called the '*representative point*'. At a later time t_1 we measure (p_1, q_1), another representative point, and the line joining (p_0, q_0) to (p_1, q_1) is the trajectory t_0–t_1 in *phase space*. The real trajectory is, of course, a motion along the q axis and is quite different.

We now return to the ensemble idea. If we measure the (p, q) values for a total of n oscillators making up the ensemble, we can obtain a representative point locating each oscillator in phase space. If, by some curious chance, all the oscillators had nearly the same values of p and q, the representative points would be bunched together around these values. Such an observation might be made if all the oscillators were set going, in phase, at time t_0. However, the interaction energy will act so as to randomize the phases and, if the observation period is taken well after the start of the experiment, we should imagine that the representative points would be more or less evenly distributed over the shaded area. We shall later show the importance of this even distribution of representative points as a fundamental postulate of statistical mechanics.

We have so far used two-dimensional space as an illustrative example. In generalized, multi-dimensional space it is usual to write the volume element of phase space* as

$$d\Gamma = dq_1 \, dq_2 \, dq_3 \ldots dq_{3N} \cdot dp_1 \, dp_2 \, dp_3 \ldots dp_{3N} \qquad 3.5$$

In strictly classical statistical mechanics there is no *a priori* limitation on the number of representative points which may be found in unit volume of phase space. In other terminology, the one we shall use, the *density* of representative points in phase space is unlimited. However, we have nowadays heard about quantum effects and it is more realistic, as well as much simpler in the long run, to recognize from the start that the uncertainty principle prevents us from dividing the phase space into infinitessimal volumes: indeed, for a single degree of freedom the uncertainty principle states that $\Delta p \, . \, \Delta q > 0(h)$, where h is Planck's constant. We can therefore state that, if $f = 3N$ degrees of freedom are involved, a volume element $d\Gamma$ cannot contain more than $d\Gamma/h^f$ representative points. Transforming this to energy, the maximum number of points is

$$\frac{1}{h^f} \cdot \frac{\partial \Gamma}{\partial E} \cdot \Delta E \qquad 3.6$$

In classical statistics we shall be concerned with the *probability* of finding a point in $d\Gamma$, which probability is directly proportional to $d\Gamma/\Gamma$, and therefore the h^{-f} factors cancel. The quantum viewpoint removes a tiresome and inessential worry about limits. In quantum statistics, h will reappear explicitly in the final results, as it ought to.

A few comments on the geometry of multi-dimensional space are now in order. The concept of 'sphere' is retained and an object which is called a 'hypersphere' is introduced. The radius of a $6N$-dimensional hypersphere is

$$R = \{p_1^2 + p_2^2 + p_3^2 + \ldots + p_{3N}^2 + q_1^2 + q_2^2 + q_3^2 \ldots . q_{3N}^2\}^{\frac{1}{2}} \qquad 3.7$$

*The volume element of phase space for a *single* free particle is six-dimensional and is often written $d\mu$. Thus, 'Γ-space' and μ-space' are terms often used in the literature.

and the volume is

$$V = \frac{\pi^{3N} R^{6N}}{\Gamma(3N + 1)} \qquad 3.8$$

In equation 3.8, $\Gamma(3N + 1)$ is the gamma function of argument $3N + 1$, and we have (temporarily) used V for volume instead of Γ. The reader can check that equation 3.8 is correct for ordinary space. Since the total energy, the Hamiltonian, is a quadratic function of p's and q's, equation 3.7 shows that the radius of the hypersphere is proportional to the square root of the energy. Since the *same* energy can be achieved in a very large number of different ways, the energy surface is covered by a large number of representative points. If the phase space is being used to represent gas molecules, for example, the translational ke can vary from 0 to ∞ (strictly, to $\frac{1}{2} Mc^2$, according to relativity theory, but this is so large that we can think of an indefinitely large limit); the hypersphere will then have a very large radius and representative points will be found at all radii inside it. If we consider a series of spherical energy shells, corresponding to equal increments of energy ΔE, or of radius ΔR, then it is clear that, from the form of equation 3.8, there are enormously more representative points available in the outer shells than in the inner ones. However, we know that the average energy of a gas molecule, the thermal energy, is quite low and we can conclude that most of the outermost shells are void of points while the inner ones are more densely populated. The probability of finding an allowed point occupied is then far from constant over the hypersphere and must be high for low energies and very low for high energies. Explicitly we can write

$$\text{number of available representative points} = \frac{\text{const.}\, E^{3N}}{h^{3N}} \qquad 3.9$$

where E is the maximum energy allowed by the problem. Since in statistical mechanics $3N = 10^{10}$ would be a rather small value, equation 3.9 gives enormous numbers unless $E \leqslant h$.

In quantum statistics, the energy levels are frequently discrete rather than continuous, but, as the energy increases, the spacing between levels becomes trifling compared with the energy itself and we can move over to a continuum representation. The discrete levels are envisaged as shells and the remarks made above can easily be adjusted to cover this case.

We can dispose of velocity (or momentum) space and configuration space rather rapidly. Velocity space is the hyperspace in which the $3N$ velocity or momentum ($\mathbf{p} = m\mathbf{v}$) components form the axes. Velocity space is frequently used in problem solving. In configuration space, $3N$ coordinates formthe axes and this space is much more rarely useful than phase space or velocity space.

3.2 The Gibbs ensembles

In his formulation of statistical mechanics, Gibbs faced the problem of introducing a more abstract theory which would cover assemblages of particles much more complicated than gas molecules – in particular, macroscopic objects. We have already said that he did this by visualizing 'ensembles' – collections of very large numbers of identical systems, which could be microscopic *or* macroscopic – and by probability calculations predicting the most probable behaviour of the ensemble. He then assumed that the behaviour of the ensemble was the same as the long-time average behaviour of a single system of the type under consideration. To carry out his calculations, he introduced three standard, or canonical, ensembles to which real experiments could be approximated. Before doing any mathematics, we shall give a brief discussion of the physical ideas behind these ensembles. This is particularly relevant because the names Gibbs gave his ensembles are, to some extent, misleading as to the role they play.

First, Gibbs defined the 'microcanonical ensemble'. Physically, the microcanonical ensemble can be looked on as a large heat bath, very carefully insulated from the surrounding objects so that the temperature (and therefore the total energy) is held as nearly constant as possible. It is better not to think of the bath as thermostatically controlled, because any real thermostat will introduce a periodic change of temperature, although a very good thermostat will involve only a small change. A rather better visualization would be that the lagging is made as good as possible, the rate of decrease of temperature is measured, and, to compensate, a small uniform supply of heat is fed into the bath to oppose the measured loss. Then, the temperature and the mean energy of the microcanonical ensemble are known to within a very small range, say ΔE_0 at E_0. ΔE_0 is, however, a macroscopic range and not a quantum range and must be visualized as a representation of a Dirac delta function rather than as a mathematical delta function. The ensemble inside the heat bath can then be thought of as showing a very small range of energies. Coupling between the systems of the ensemble is considered to be reduced to a minimum but, inevitably, cannot be zero, so if we inspect a particular system it will have energy in a certain range ΔE_1 at E_1. We shall see later that this really characterizes the microcanonical ensemble. A consequence is that we can do little mathematically with this ensemble; however, its conceptual value is very great.

Somewhat surprisingly, the 'canonical' ensemble ought to be thought of as an ensemble contained *within* the microcanonical ensemble, for it is an ensemble which is in contact *only* with a microcanonical ensemble and is otherwise completely isolated from the rest of the world. The container which separates the canonical from the microcanonical ensemble is such that it allows the exchange of heat between the two ensembles but *not* the exchange of particles or systems. Thus, the number of systems or microparticles in the canonical ensemble is constant, while the energy of the whole heat bath, but not that of the (small) volume representing the canonical ensemble, remains constant. We shall see that we can apply mathematical techniques to obtain a great deal of information about the canonical ensemble.

Lastly, Gibbs defined the 'grand canonical' ensemble. Here, the situation is exactly the same as in the canonical ensemble except that the wall separating the grand canonical ensemble from the microcanonical ensemble allows the passage of both heat and particles or, and this is very important, absorbs particles. In either case the number of particles in the ensemble can change. From the mathematical viewpoint, the grand canonical ensemble is the most useful and flexible of the three.

The reader will, no doubt, now appreciate the reasons for thinking that Gibbs' nomenclature is less than transparent. A quotation from Landau and Lifschitz[4] (p. 103) presses home the point.

> The description of a system by means of the microcanonical distribution is equivalent to neglecting the fluctuations of its total energy; the canonical distribution takes into account these fluctuations. The later form, in turn,

neglects the fluctuations in the number of particles and may be said to be microcanonical with respect to the number of particles. The grand canonical ensemble is 'canonical' both with respect to the energy and to the number of particles. Thus, all three distributions . . . are, in principle, suitable for determining the thermodynamic properties of the body. The only difference from this point of view lies in the degree of mathematical convenience. In practice, the microcanonical distribution is the least convenient and is never used for this purpose. The Gibbs distribution for a variable number of particles is usually the most convenient.

We now discuss these ensembles and the probability distributions which result from them, in turn.

3.3 The Gibbs microcanonical ensemble

To introduce the idea of the microcanonical ensemble, consider the ensemble which consists of systems which are isolated from the rest of the world. More exactly, we suppose an ensemble of adiabatic systems has been set up. Of course, from what has been said above, we cannot achieve complete isolation or a completely adiabatic system; we can only achieve conditions in which the interaction energy ΔE is very small by any comparison with all the other energies. We also imagine that the energies of all the elements of the adiabatic system are in the range $E_0 \to E_0 + \Delta E_1$. Then we can proceed by two routes.

We can argue that, because of the existence of ΔE_1 even when very small, the members of the ensemble will be present with all possible values of the p's and q's. Revert, for the moment, to the harmonic oscillator. The phase of the individual oscillators is the argument of $\begin{smallmatrix}\sin\\ \cos\end{smallmatrix}(\omega t + \phi)$, but, owing to the interaction, ω and ϕ are randomly distributed over a small range. Thus, in equilibrium when a long time has elapsed after preparation of the ensemble, $\begin{smallmatrix}\sin\\ \cos\end{smallmatrix}(\omega t + \phi)$ can have any value between ± 1, and the (p, q) values vary accordingly and will uniformly fill the elliptical energy shell of Fig. 3.1.

Alternatively, according to Liouville's theorem of classical dynamics,[1,2,3] the density of representative points is invariant with time and this can be true only if the shell is uniformly filled. Liouville's theorem is proved in references 1 to 3 for macroscopic and microscopic particles, which do *not* interact. It is therefore satisfactory for representative points to the extent that these are non-interactive, but there are conceptual difficulties when even weak interactions are allowed. Thus, there is less clear advantage to be gained by using Liouville's than might be thought. Either way, we consider that the density of representative points is constant over the energy shell. Alternatively, the probability of finding a system in a small fixed volume of the energy shell is a constant which depends only on the energy.

So far, the implication has been that we are considering classical systems and macroscopic motions. In quantum statistics, the system is a microscopic particle: an atom or a molecule. In such cases, we must remember that, in addition to the energy due to macroscopic motion, it will usually be more important to consider the quantized energy structure of the microparticle itself. Each energy state of the system is specified by a set of f quantum numbers if there are f degrees of freedom. Reference to elementary solutions of the Schrödinger equation shows that, in these cases, a definite value of energy is associated with each solution and these are described by a set of quantum numbers. As an example, we may quote the case of an electron in a potential box of sides x_1, y_1, z_1, where the energy is

$$E_{n,\,m,\,l} = \text{const.}\,\hbar^2\left(\frac{n^2}{x_1^{\,2}} + \frac{m^2}{y_1^{\,2}} + \frac{l^2}{z_1^{\,2}}\right)$$

where n, m, l are the quantum numbers. If E is such that n, m, l are large numbers, we note that the energy difference $E_{n+1,\,m,\,l} - E_{n,\,m,\,l}$ can be small and in the range ΔE_1 at E, so that transitions between quantum states can occur for small interactive energies. Moreover, we must also recognize the existence of energy degeneracy. This means that two (or more) different quantum states exhibit the same value of energy E, and this clearly involves $(n^2/x_1^{\,2} + m^2/y_1^{\,2} + l^2/z_1^{\,2}) = \text{const.}$, an expression which will yield several different sets of (n, m, l) at least for large energies. Thus, changes of state without change of energy are also possible. In an ensemble of microparticles we therefore conclude that, no matter how small the interaction energy ΔE_i, vast numbers of states will still be accessible to the particles of the ensemble. Since there is no reason why a particular state with a certain energy should be more likely than another state with the identical energy, we must conclude that the probability of finding a particular system (microparticle) in a particular state is a constant, provided the state is one which has an energy in ΔE_i at E.

If we, from now on, denote probabilities by the symbol W (we have used p for momenta), we can now write down a mathematical expression for the probability. Let $\Omega\,(E)\mathrm{d}E$ be the number of quantum states or representation points in $\mathrm{d}E$ at E, then Ω clearly is a function of E and also depends on the system, i.e. it depends on whether the system is a microparticle in a box or whether it is a microparticle oscillating on a line etc. Then, on the basis of our reasoning so far, we may write

$$W(E_i) = \text{const.}\ \Omega(E_i) \qquad 3.10$$

The constant in equation 3.10 is to be determined by the condition that

$$\sum_{i=1}^{\infty} W(E_i) = 1.0$$

In words, the probability of finding a particle in a state with a fixed energy is directly proportional to the number of states with that energy. Equation 3.10 is the mathematical expression of Gibbs' *microcanonical ensemble.* It is much less trivial than it might appear, but this will emerge as we proceed.

3.4 The canonical ensemble

Let us carry straight on with the canonical ensemble. here, we visualize a very large *isolated* system, which is usually called the 'reservoir', in which a much smaller system is immersed, but which still involves an enormous number of degrees of freedom. An example might be a small hot metal object dropped into a large tank of cooling water. We ask ourselves, what is the probability of finding the small system with a specified energy value? It is important to remember that the probability found relates only to the complete system of cooling water and hot metal *after* equilibrium has been established; the transient process is much more difficult and is considered later.

Let E_k = energy of reservoir in the kth state
E_i = energy of object in the ith state

Then, because the isolation conserves energy,

$$E = E_k + E_i = \text{const.} + \delta E$$

Here δE is the interaction energy, which we have earlier said cannot be identically zero. But, under our present assumptions (large reservoir) $\delta E \ll E_i \ll E_k$ so δE is negligible from a numerical viewpoint. After equilibrium is reached, the ensemble as a whole is isolated and can be considered as a microcanonical ensemble. Thus, using equation 3.10, we can write

$$W_i = C\Omega(E) = C\Omega(E_k + E_i) \tag{3.11}$$

The meaning of W_i, it is stressed, is the probability that the state is one in which the energy of the object is E_i and that of the reservoir is simultaneously E_k. But the numbers of states $\Omega(E_k + E_i)$ must equal the product $\Omega(E_k) \,.\, \Omega(E_i)$, according to the law of multiplication of probabilities of independent events, or, which is precisely equivalent from equation 3.11,

$$\Omega(E_k + E_i) = \Omega(E - E_i) \,.\, \Omega(E_i)$$

then $$W_i = C \,.\, \Omega(E - E_i) \,.\, \Omega(E_i) \tag{3.12}$$

The next step is to expand $\Omega(E - E_i)$ as a power series in the small decrement E_i. Here we must make multiplication of numbers of states correspond with addition of energy terms.* The natural logarithm has this property, so we make the expansion on $\ln \Omega$ instead of directly on Ω:

$$\ln \Omega(E - E_i) = \ln \Omega(E) - [\partial \ln(\Omega)/\partial E]_0 E_i + \text{etc.} \tag{3.13}$$

Call the square bracket β, then

$$\ln \Omega(E - E_i) = \ln \Omega(E) - \beta E_i$$

$$\therefore \ \Omega(E - E_i) = \text{const.} \times \exp(-\beta E_i) \tag{3.14}$$

or $$W_i = C' \exp(-\beta E_i)\, \Omega(E_i) \tag{3.15}$$

By definition, we must put

$$C' = \frac{1}{\sum_i \exp(-\beta E_i)\, \Omega(E_i)} \tag{3.16}$$

*Making the radius of the hypersphere a little greater is equivalent to multiplying the surface by an amount proportional to the energy increment to ensure that $\Sigma W_i = 1.0$ and to reach the final result

$$W_i = \frac{\exp(-\beta E_i)\, \Omega(E_i)}{\sum_i \exp(-\beta E_i)\, \Omega(E_i)} \tag{3.17}$$

Equation 3.17 defines the probability distribution which is called the 'Gibbs distribution' or the 'canonical (standard) distribution'. Equation 3.17 is often rewritten in the economical form

$$W_i = \frac{\exp(-\beta E_i)\, \Omega(E_i)}{Z} \tag{3.18}$$

where $Z = \sum_i \exp(-\beta E_i)\, \Omega(E_i)$ is called the 'partition function'. In their present form, equations 3.17 and 3.18 are not very informative: they merely state that the probability depends on the product of a decreasing exponential and $\Omega(E_i)$, which certainly increases rapidly with E_i in the case of the particle in a box. To anticipate, we shortly show that

$$\beta = 1/kT$$

where k = Boltzmann's constant and T = statistical temperature.

Therefore, if $E_i \gg kT$ which is not difficult for laboratory temperatures, W_i is likely to be very sharply peaked over a particular small range of energies, as shown

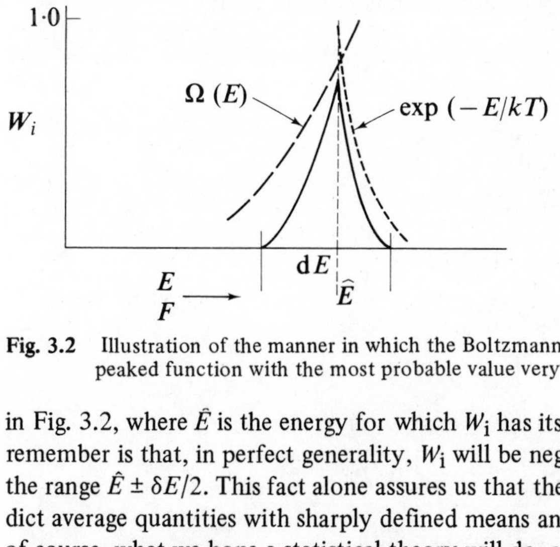

Fig. 3.2 Illustration of the manner in which the Boltzmann theorem leads to a sharply peaked function with the most probable value very nearly equal to the mean value.

in Fig. 3.2, where $\hat{E}$ is the energy for which W_i has its maximum value. The point to remember is that, in perfect generality, W_i will be negligible when E_i lies outside the range $\hat{E} \pm \delta E/2$. This fact alone assures us that the Gibbs distribution will predict average quantities with sharply defined means and small fluctuations. This is, of course, what we hope a statistical theory will do, and is what it must do if it is to have any hope of agreeing with experiment.

3.5 The mean energy and energy dispersion in a Gibbsian distribution

From equation 3.18, the mean energy $\langle E\rangle$ is

$$\langle E\rangle = \frac{\sum_i \exp(-\beta E_i)\, \Omega(E_i)\, E_i}{Z} \tag{3.19}$$

Inspection shows that the numerator is the negative derivative of the denominator, taken with respect to β. Then

$$\langle E\rangle = -\frac{1}{Z}\frac{\partial Z}{\partial \beta} = -\frac{\partial}{\partial \beta}(\ln Z) \qquad 3.20$$

The energy dispersion $\langle \Delta E^2\rangle$ is defined by

$$\langle (E-\langle E\rangle)^2\rangle = \langle E^2\rangle - \langle E\rangle^2$$

But $$\langle E^2\rangle = \sum_i \exp(-\beta E_i)\,\Omega(E_i)\,E_i^2$$

$$= \frac{1}{Z}\frac{\partial^2 Z}{\partial \beta^2} \qquad 3.21$$

Therefore $$\langle \Delta E^2\rangle = \frac{1}{Z}\frac{\partial^2 Z}{\partial \beta^2} - \frac{1}{Z^2}\left(\frac{\partial Z}{\partial \beta}\right)^2 = \frac{\partial^2(\ln Z)}{\partial \beta^2} \qquad 3.22$$

These formulae illustrate the importance of the partition function.

Example

A very easy example of the case of the partition function is the simple harmonic oscillator. In quantum theory the energy eigenvalues are $E_i = (i + \frac{1}{2})\,\hbar\omega$, $i = 0, 1, 2, \ldots, \infty$. The degeneracy $\Omega(E_i)$ by inspection is $\Omega = 1.0$. Then

$$Z = \sum_i^{\infty} \exp\left\{-(i+\tfrac{1}{2})\hbar\omega\beta\right\}$$

$$= \exp -\frac{\hbar\omega\beta}{2}\cdot\sum_i \exp(-i\hbar\omega\beta)$$

$$= \exp\left(-\frac{x}{2}\right)\cdot\frac{1}{1-\exp(-x)} \qquad (x = \hbar\omega\beta)$$

$$= \frac{1}{\exp(+x/2)+\exp(-x/2)} \qquad 3.23$$

therefore $$-\frac{\partial Z}{\partial \beta} = \frac{\hbar\omega}{2}\cdot\frac{\exp(+x/2)+\exp(-x/2)}{[\exp(+x/2)-\exp(-x/2)]^2}$$

and $$\langle E\rangle = \frac{\hbar\omega}{2}\cdot\frac{\exp(+x/2)+\exp(-x/2)}{\exp(+x/2)-\exp(-x/2)}$$

$$= \frac{\hbar\omega}{2} + \frac{\hbar\omega}{\exp\beta\hbar\omega - 1} \qquad 3.24$$

If we identify β as $1/kT$ and subtract the zero-point energy $\hbar\omega/2$, this is identical with Planck's (intuitive) hypothesis. We shall later rederive equation 3.24 using Bose–Einstein statistics. The development here shows two things: the generality of

Gibbs' method and a strong incentive to put $\beta = 1/kT$, which really ought to be proved thermodynamically – this we shall shortly do.

3.6 The grand canonical ensemble

For obvious reasons, the grand canonical ensemble is often called the 'Gibbs ensemble for a variable number of particles'. Though long-winded, this term has the merit of clarity. It may be helpful to think of a pair of simple systems which obey this kind of ensemble. One example, of interest in chemistry, is that of a vapour in equilibrium with the parent-liquid surface; another, more interesting to us, is electromagnetic radiations. In quantum theory the e.m. waves become photon beams. When such beams collide with walls of an enclosure (cavity resonator) they are absorbed so that the number of photons is *not* conserved.

We must then describe a system, of constant volume, which can exchange both heat and particles with the surrounding heat bath. If the problem were merely that the number of particles in the volume V were not constant, it would be readily soluble. However, we must remember instead that the number of energy levels available to the system is a function of the number of particles included in the system at any instant. We therefore use the following notations: let E_{nN} be the energy levels, depending on the quantum numbers or generalized coordinates n and also on N, the number of particles at a given time (i.e. N is the instantaneous number of particles in V). The total number of particles present inside the reservoir, or microcanonical ensemble, is constant, symbol N_0, and E_R is the large, constant energy of the reservoir and N_R is the number of particles in the reservoir. Then

$$N_0 = N + N_R$$

The treatment is now similar to that of section 3.4. Equation 3.12 is transformed to

$$W_{nN} = C\,\Omega(E_R - E_{nN}, N_0 - N)\,.\,\Omega(E_{nN}, N) \qquad 3.25$$

We expand $\ln \Omega(E_R - E_{nN}, N_0 - N)$ as before, using the argument about the relative smallness of the sub-system to justify the retention of only the first terms of the expansion. The result is

$$W_{nN} = C \exp[\Omega(E_R, N_0) - \beta E_{nN} + \mu\beta N]\ \Omega(E_{nN})$$

where $\beta = [\partial(\ln \Omega)/\partial E]_{E=0}$ as before 3.26

$$\mu\beta = -[\partial(\ln \Omega)/\partial N]_{N=N_0} \qquad 3.27$$

Therefore $W_{nN} = C' \exp\{-\beta(E_{nN} - \mu N)\}\Omega(E_{nN}, N)$ 3.28

Once again, C' is determined from $\Sigma\Sigma W_{nN} = 1.0$,

i.e. $$C' = \frac{1}{\sum_n \sum_N \exp\{-\beta(E_{nN} - \mu N)\}\,\Omega(E_{nN}, N)} \qquad 3.29$$

Inserting this into the expression for W_{nN}, we obtain the grand canonical distribution. If N = const., μ is zero from equation 3.27 and the distribution reduces correctly to the canonical distribution function.

The grand partition function is defined slightly differently from the way one might expect. Call it Z_g; then*

$$Z_g = \exp(-\mu\beta N) \sum_n \sum_N \exp\{-\beta(E_{nN} - \mu N)\} \Omega(E_{nN}; N) \qquad 3.30$$

and $$W_{nN} = \frac{\exp(-\beta E_{nN}) \Omega(E_{nN}, N)}{Z_g} \qquad 3.31$$

In using these results, it must be remembered that the volume element in phase space itself depends on the number of particles N, since there are $6N$ degrees of freedom. To make this explicit, we write

$$\Omega(E_{nN}, N) = \frac{\Delta q_1 \ldots \Delta q_{3N} \cdot \Delta p_1 \ldots \Delta p_{3N}}{h^{3N}} = \frac{\partial \Gamma_n}{h^{3N}}$$

so, finally, $$W_{nN} = \frac{\exp(-\beta E_{nN})\, d\Gamma_n}{Z_g \,.\, h^{3N}} \qquad 3.32$$

which is the probability of finding the sub-system in the energy state or level n with, simultaneously, exactly N particles in it.

Before discussing the canonical and grand canonical distributions further, we shall consider another method of deriving them, which is very interesting and has been widely used by Schrödinger[5].

3.7 Another derivation of Gibbs' distribution

Schrödinger's derivation starts off from a rather different standpoint from that of Gibbs; the idea is quite simple, however. The microcanonical ensemble is again used as a heat reservoir. Consider that the system described by Gibbs has inside it a very large number of sub-systems, each itself large enough to comprise a large number of macro- or microparticles and each particle having a large number of energy eigenvalues or, in the continuous case, a wide energy spectrum. In other words, each subsystem has a very large number of degrees of freedom, and the sub-systems can exchange energy and particles among themselves and can exchange energy, but *not* particles, with the reservoir. Then, if the total number of particles in all the subsystems is N, the number of particles in each sub-system, a_1, a_2, a_3, etc., fluctuates but the equality $N = \sum_i a_1$ is always obeyed.

Examined closely, this description relies on the distinguishability of microscopic particles. Since we now realize that microparticles are not, in fact, distinguishable, it does not really apply to some, or even most, classical situations. A changed viewpoint which relies on indistinguishable particles is the following.

*When using equation 3.31 to form averages, $\langle N \rangle$ for example, remember the factor $\exp(-\mu\beta N)$ in Z_g.

The division into sub-systems given above is quite arbitrary – a thought experiment, in fact – and another division is equally valid. For this, we consider that each sub-system is characterized by an energy eigen value and that the state with energy E_l has an instantaneous occupation number a_l. We now have a very large number of sub-systems, each of which has an allowed energy state, labelled E_1, E_2, etc., ordered so that $E_{l+1} \geqslant E_l$. The total energy is conserved and so is the total number of particles N. The sub-systems (states) of the ensemble can be tabulated as follows:

State number	1	2	3	$\ldots l$
Energy	E_1	E_2	E_3	$\ldots E_l$
Occupation no.	a_1	a_2	a_3	$\ldots a_l$

For the moment, we can consider the a's as fixed numbers; later we shall vary them consistently with the number conservation. This implies an energy variation which must obviously be consistent with energy conservation over the system. The problem is basically a combinatorial one. We find an expression for the number of different ways in which the table can be realized, and maximize this result, subject to the two conservation rules. This determines the most probable arrangement. It is then easy to show that the most probable arrangement is overwhelmingly more probable than any other arrangement; therefore, the most probable arrangement is also the arrangement which, on average, will be found in a measurement.

The number of different ways in which the tabulated arrangement can be reached is

$$P = \frac{N!}{a_1!\, a_2! \ldots a_l!} \qquad 3.33$$

which can be derived by writing down the number of ways in which we can put a_1 objects in box no. 1, a_2 in box no. 2, etc., when we have a total of N objects. There are

$$\frac{N!}{a_1!\,(N-a_1)!} \quad \text{ways of filling box no. 1,}$$

$$\frac{(N-a_1)!}{a_2!\,(N-a_1-a_2)!} \quad \text{ways of filling box no. 2, and so on.}$$

Multiplying the results together, cancelling the common terms and remembering that $0! = 1$, we reach equation 3.33.

It is far easier to maximize $\ln P$ than P itself, and the maximization is subject to the two constraints:

$$\sum_l a_l = N = \text{const.} \qquad 3.34$$

$$\sum_l a_l E_l = E = \text{const.} \qquad 3.35$$

The maximization is performed by using Lagrangian multipliers, μ for equation 3.34 and β for equation 3.35. We then look for the maximum of

$$\ln P - \mu \sum_l a_l - \beta \sum_l a_l E_l \qquad 3.36$$

To get rid of the factorials in P, we can use the Stirling approximation in the simplified form

$$\ln x! \approx x \ln x - x \qquad 3.37$$

This is valid for large x, $> 0(100)$, so that it is restricted in validity to the a_l's which are large. We may, however, argue that the contribution to the probability made by improbable states is likely to be very small, so that an inaccurate estimate of this contribution will not make any difference to the final result. Then, treating the variation of the a's as though they were continuous variables, we find, after a little algebra, the maximum condition is

$$-\sum_i^{\infty} (\ln a_l + \mu + \beta E_l)\, \delta a_l = 0 \qquad 3.38$$

To resolve this equation we first fix μ and β by requiring that

$$\ln a_1 + \mu + \beta E_1 = 0$$

$$\ln a_2 + \mu + \beta E_2 = 0$$

then $$\sum_{l=3}^{\infty} (\ln a_l + \mu + \beta E_l) = 0$$

which can be true for fixed μ and β only if each term of the series is itself zero. Thus, we have proved that, for all l,

$$\ln a_l + \mu + \beta E_l = 0 \qquad 3.39$$

Equation 3.39 is equivalent to $a_l = \exp[-(\mu + \beta E_l)]$. 3.40

Equation 3.40 can then be put back into equations 3.34 and 3.35 to find

$$N = \sum_l \exp[-(\mu + \beta E_l)] \qquad 3.41$$

$$E = \sum_l E_l \exp[-(\mu + \beta E_l)] \qquad 3.42$$

Expressing these differently,

$$\exp(-\mu) = \frac{N}{\sum_l \exp(-\beta E_l)} \qquad 3.43$$

or $$a_l = \frac{N \exp(-\beta E_l)}{\sum_l \exp(-\beta E_l)} \qquad 3.44$$

it being emphasized that μ and β are constants which still have to be given physical significance. We can now write down the average energy per particle because, by definition, it is merely E/N.

$$\text{Therefore} \quad \langle E \rangle = \frac{\sum_l E_l \,.\, \exp(-\beta E_l)}{\sum_l \exp(-\beta E_l)} \tag{3.45}$$

using equations 3.41 and 3.42.

Let us now compare these results with those we found in sections 3.4 and 3.6. First, in the above treatment we have implied that the energy eigenvalues are unique or, in the technical terminology, 'non-degenerate'. If each energy level can be described by several different sets of quantum numbers, it is called 'degenerate' and one can define a degeneracy factor g_l which indicates the number of states with the common eigenvalue E_l. This factor g_l can be regarded as a (known) statistical weighting factor. If we *had* included it from the start, equation 3.45 would appear in the modified form

$$\langle E \rangle = \frac{\sum_l g_l \,.\, E_l \exp(-\beta E_l)}{\sum_l \exp(-\beta E_l)} \tag{3.46}$$

Alternatively, from equation 3.44 we can write down the probability of a state l being occupied, which is a_l/N, as

$$W_l = \frac{g_l \exp(-\beta E_l)}{\sum_l g_l \exp(-\beta E_l)} \tag{3.47}$$

If we identify g_l with $\Omega(E_i)$ and assume that β has the same meaning in both equations, equation 3.47 is identical with equation 3.17. This means that we have retrieved the Gibbs canonical distribution. On the other hand, in reaching equation 3.47 we have eliminated μ, the second Lagrange multiplier which is presumably just as significant as β. In section 3.6 we saw that the Gibbs grand canonical ensemble involved two constants which we defined by equations 3.26 and 3.27. Therefore, if we suppose that μ defined by equation 3.27 is, or can be made, the same as μ defined by equation 3.43, we have also achieved a representation of the grand canonical distribution.

It is clearly time to discuss the physical identity of μ and β, but before doing so we ought to dispose of the slightly unsatisfactory mathematics of the derivation we have just discussed. Many writers have found this derivation unsatisfactory, although it is well known to lead to results which are certainly correct. The problem has been tackled in two quite different ways. The first, due to Darwin and Fowler[6,7,8] uses a different mathematical technique, that of steepest descents, to solve for the most probable distribution. Today, this method is held to be mathematically rigorous and to prove the results cited; however, it is mathematically very formidable and is somewhat outside the mainstream of modern thinking on these problems. The second approach, due to Khinchin[9], exploits the limit theorems of modern probability theory to reach rigorous proofs of the required formulae. Khinchin's work is advanced but not unduly difficult and has the advantage of always keeping firmly in view the kind of mathematical thinking used in modern kinetic theory (which we introduce in later chapters). While a decision between these two methods may be largely a matter of personal taste, the fact is that the fundamental formulae of statistical mechanics can be given a fully rigorous derivation and are not to be considered as semi-empirical in nature.

3.8 The statistical temperature

The meaning of β can be established in several different ways; for example, one can apply the development to gas molecules and rederive the Maxwell distribution, when it will be necessary to put $\beta = 1/kT$. Here, we try to keep the development more general. We first stress that the property described by β is something which all the systems, sub-systems and microcanonical ensemble share in common.

Consider two separate sub-systems, described, following equation 3.18, by

$$W_1 = \frac{\exp(-\beta E_1)\,\Omega(E_1)}{Z} \qquad W_2 = \frac{\exp(-\beta E_2)\,\Omega(E_2)}{Z}$$

These formulae show that β is fundamentally a *positive* number because it is impossible to believe that the probability of finding a particle in a given state is an *increasing* function of the energy of the state.

To make our argument in the simplest way, define $\Theta_1 = 1/\beta_1, \Theta_2 = 1/\beta_2$ and allow our two systems to come into a state of weak coupling. By the law of multiplication of probabilities, this state must have probability

$$W = W_1 W_2 = \frac{\exp(-E_1/\theta_1)\,\Omega(E_1)\exp(-E_2/\theta_2)\,\Omega(E_2)}{Z^2} \qquad 3.48$$

But, in allowing the sub-systems to interact weakly, we have in no way violated the assumptions of the original Gibbs canonical ensemble and we must therefore also be able to express W as

$$W = \frac{\exp(-E/\theta)\,\Omega(E)}{Z} \qquad 3.49$$

which can be the case only if $\theta_1 = \theta_2 = \theta$. We know very well that, if we make the experiment, the property which reaches equilibrium in this way is the thermodynamic temperature; therefore, we can identify the parameter θ as the statistical temperature. It is related to the thermodynamic temperature by

$$\theta = kT \qquad 3.50$$

where Boltzmann's constant, k, is merely a number needed to express θ in kelvin. Therefore

$$\beta = 1/kT \qquad 3.51$$

We must defer making the identification of μ until we have discussed some results from statistical thermodynamics. Instead, let us continue with an application of the results reached so far.

3.9 Gibbs' distribution for an ideal gas

The physical system considered here is a box, volume V, in which we know there are N particles of an ideal gas. The system is at constant temperature T and we assume that it was prepared long enough before the start of observations for equilibrium to have been reached. In equilibrium, the probability of finding a particle (or molecule) with a definite translational kinetic energy does not depend on the coordinates, but only on the energy. Since we are dealing with a classical situation, the energy is a continuous function and we can write the partition function as an integral rather than as a sum over discrete states. It is

$$Z = \int \exp(-\beta E)\,\Omega(E)\,\mathrm{d}E$$

The number of states in $\mathrm{d}E$ at E is, as discussed earlier,

$$\frac{1}{h^{3N}} \cdot \frac{\partial \Gamma}{\partial E} \cdot \mathrm{d}E$$

therefore $$Z = \frac{1}{h^{3N}} \int \exp(-\beta E) \frac{\partial \Gamma}{\partial E} \cdot \mathrm{d}E \qquad 3.52$$

Also, we can write equation 3.18 in the forms

$$\mathrm{d}W = \frac{\exp(-\beta E)}{h^{3N} Z} \cdot \frac{\partial \Gamma}{\partial E} \cdot \mathrm{d}E \qquad 3.53(a)$$

$$= \frac{\exp(-\beta E)\,\mathrm{d}\Gamma}{h^{3N} Z} \qquad 3.53(b)$$

$$= \frac{\exp(-\beta E)\,\mathrm{d}\Gamma}{\int \exp(-\beta E)\,\mathrm{d}\Gamma} \qquad 3.53(c)$$

from which the quantum factor h^{3N} has been eliminated. Equation 3.53, which is the classical Gibbs distribution, therefore agrees with the correspondence principle that quantum relationships → classical ones as $h \to 0$.

We now apply this to our problem. The element of $6N$-dimensional space is

$$\mathrm{d}\Gamma_{6N} = \mathrm{d}p_1\,\mathrm{d}p_2 \ldots \mathrm{d}q_1\,\mathrm{d}q_2 \ldots \mathrm{d}q_{3N}$$

and $$E = \frac{1}{2m} \sum_k p_k^{\,2} \qquad 3.54$$

Then $$\Gamma = \int_{6N} \mathrm{d}\Gamma = V^N \int_{3N} \mathrm{d}p_k \qquad 3.55$$

since $\mathrm{d}V = \mathrm{d}q_1\,\mathrm{d}q_2 \ldots \mathrm{d}q_3$

Consider for a moment the special case $N = 1$; then

$$E = \frac{1}{2m}(p_1{}^2 + p_2{}^2 + p_3{}^2) = \frac{\hat{p}^2}{2m}$$

or, in other words, the p-space can be represented by a sphere of radius $\sqrt{(2\,mE)}$

whose volumé is $V_p = (4\pi/3)\,(2mE)^{\frac{3}{2}}$

For this case, then, equation 3.55 gives

$$\Gamma = V \,.\, V_p = \frac{4\pi V}{3} \,.\, (2mE)^{\frac{3}{2}}$$

Similarly, in $3N$-dimensional velocity space the volume corresponding to 'radius' $(2mE)^{\frac{1}{2}}$ is const. x $E^{3N/2}$. Thus,

$$\Gamma \propto V^N \,.\, E^{3N/2}$$

$$d\Gamma/dE \propto V^N E^{3N/2-1} \qquad 3.56$$

Putting this into equation 3.53(a) we get

$$dW = \frac{\exp(-\beta E)\, E^{3N/2-1}\, dE}{\int \exp(-\beta E)\, E^{3N/2-1}\, dE} \qquad 3.57$$

It should be remembered, in using equation 3.57, that the bottom line is *not* the partition function.

We can now calculate the mean energy of the N particles, as usual.

From equation 3.57, $$\langle E\rangle = \frac{\int_0^\infty \exp(-\beta E)\, E^{3N/2}\, dE}{\int_0^\infty \exp(-\beta E)\, E^{3N/2-1}\, dE}$$

Using the transformation $E = u^2$, this in turn becomes

$$\langle E\rangle = \frac{\int_0^\infty \exp(-\beta u^2)\, u^{3N+2}\, du}{\int_0^\infty \exp(-\beta u^2)\, u^{3N}\, du}$$

$$= \frac{3N+1}{2\beta}$$

$$\rightarrow \frac{3N}{2\beta} \qquad 3.58$$

In reaching equation 3.58 we have used standard integrals and have neglected 1 by comparison with $3N$, which is enormous. But we know very well that the average energy of a particle moving in three-dimensional space is $3kT/2$; therefore we must put

$$\langle E\rangle/N = 3/2\beta = 3kT/2$$

Thus $\beta = 1/kT$ 3.59

and our earlier hypothesis is verified. The more general and much more subtle thermodynamic proof leads, as it must, to the same conclusion. In future, we shall often continue to use β, to save writing. It is now defined by equation 3.59 instead of being an arbitrary multiplier.

By differentiating the numerator of equation 3.40, it may be shown that the most probable value of E is identical with the mean value of E, $\langle E\rangle$. This was implied by Fig. 3.2.

It is interesting to calculate the dispersion $\langle E^2\rangle - \langle E\rangle^2$.

$$\langle E^2\rangle = \frac{\int \exp(-\beta u^2)\, u^{3N+4}\, du}{\int \exp(-\beta u^2)\, u^{3N}\, du}$$

$$= \left(\frac{3N+3}{2\beta}\right)\left(\frac{3N+1}{2\beta}\right) \qquad 3.60$$

Therefore $$\left\{\frac{\langle E^2\rangle - \langle E\rangle^2}{\langle E\rangle^2}\right\}^{\frac{1}{2}} \approx \left(\frac{2}{3N}\right)^{\frac{1}{2}} \qquad 3.61$$

Equation 3.61 gives the standard deviation. For any normal volume of gas, N is a very large number, $10^{18} - 10^{20}$, so the standard deviation is $0(10^{-9})$ which is, of course, very small indeed. This illustrates just how sharply peaked the Gibbs distribution is in a real case – far too sharp to be represented graphically. We can approximate the Gibbs distribution by a Gaussian distribution with the specified mean (or most probable energy) and the calculated variance. We can then estimate the probability of finding, in an experiment, an average energy which deviates by any assumed amount from the calculated average. The answer is extremely small unless the specified deviation is excessively small. We shall revert to such questions when we discuss fluctuations; here, we wish to emphasize that the most probable state is *far more probable* than even closely neighbouring states. In turn, this makes the procedure of looking for the most probable state appear more attractive than it was on first sight.

3.10 Statistical thermodynamics

In this book we do not intend to dwell on the details of statistical thermodynamics (really thermostatics, since only equilibrium states are initially discussed) because many excellent textbooks cover the subject. However, to make progress in our main theme, we must demonstrate the major connections between Gibbs' statistics and thermostatic laws and derive the statistical entropy. We shall demonstrate by examples that the usual quantities defined in thermodynamics can be expressed in terms of T, the pressure, the volume and the partition function Z.

Phenomenological thermodynamics is based on the following axioms, which are verified by all presently available experimental data.

1. *The Zeroth law.* If two systems are both in thermal equilibrium with a third system at temperature T, they must also be in equilibrium with one another, at the same temperature.

2. *The first law of thermodynamics.* An equilibrium state of a system containing a large number of macroparticles has a constant internal energy E_I. If the system is coupled to another system, the change in E_I is given by

$$\Delta E_I = Q - W \qquad 3.62$$

where W = work done *by* the first system on the second
Q = the heat absorbed by the first system

The first law, in this incremental form, defines the heat Q.

3. *The second law of thermodynamics.* This can be expressed in many different ways. From our viewpoint, the requirement is to introduce the concept of entropy, which is done by the following definition. If a system is coupled to other systems and experiences a slow, quasi-static change in which it absorbs heat Q while remaining almost at constant temperature, the change in entropy is given by

$$dS = dQ/T \qquad 3.63$$

The second property of the entropy we recognize is that, if an isolated system experiences internal changes, such changes will be such that $\Delta S \geqslant 0$. The second law is often rather loosely expressed in the form, 'heat cannot flow from a colder system to a hotter system.'

4. *The third law of thermodynamics, or Nernst's heat theorem.* The entropy S has a limiting value S_0 as the temperature $T \rightarrow 0$. S_0 is independent of the nature of the particular system and, in practice, is usually taken as zero.

Of the five quantities E_I, Q, W, S and T we have a perfectly clear idea of W and T, and these are readily measurable. E_I is conceptually clear but the method for measuring it is obscure; Q is less clear and its measurement would have to emerge as a result of further relationships; and S appears a very arbitrary constant. However, as we shall shortly see, the statistical analysis results in relatively simple and clear relationships for all these quantities and, for that matter, for derived thermodynamic state variables as well. We therefore now take up the question of statistical definitions of the basic thermodynamic variables.

3.10.1 Work and internal energy in statistical terms

To understand the relationships between work and internal energy we must take the external agencies into account; that is, we must consider that, in going from one equilibrium state to another such, the system is subject to external constraints, for example the change takes place at constant pressure or constant volume or under constant temperature. We shall discuss the detailed meaning of this more deeply in a moment. First, let us make another qualification which is only implied in the above remarks. If we are to use the Gibbs distribution to describe our system, we are forced to base the discussion of any change of state on those changes which can be regarded as a very slow passage from one equilibrium state to the next. The reason for this is simply that, as yet, we have no mathematical apparatus for the study of changing or transient processes. Clearly, at this point, we cannot ascribe any definite meaning to the word 'slow'. Our later consideration of kinetic theory will actually give so-called relaxation times which provide the required numerical estimates, but these are concepts outside the Gibbs statistics. Slow changes from one Gibbsian equilibrium to another are called 'quasi-static' changes, and it is only these changes that we shall use.

Returning now to an isolated sub-system which is taken through a quasi-static change and which is subject to constraints of some sort, we find that the internal energy or total energy, defined by $\Sigma\, w_l E_l$, is subject to two kinds of change or variation. First, there are changes in which w_l is a fixed function of E but the energy levels E_l depend on the parameters and change. For example, if a cubical metal block is removed from a heat bath at one temperature T and put into a hotter one, it will expand so that a side length $x \to x + dx$. This change clearly modifies the allowed eigenvalues of the Schrödinger equation. Equally, a gas in a strong metal spherical container will change at constant volume so the pressure will increase, involving increased momentum transfer from the walls to the gas and changing the energy spectrum. Then the total energy change

$$\delta E = \sum_l \delta(E_l w_l) = \left\{\sum_l (\delta E_l \,.\, w_l)\right\}_{w_l = \text{const.}} + \left\{\sum_l (E_l \,.\, \delta w_l)\right\}_{R = \text{const.}} \qquad 3.64$$

In this expression the δ's are variational signs and R is a generalized parameter. Take the two terms on the r.h.s. in turn. We can replace $\Sigma\, \delta E_l \,.\, w_l$ by $-\Sigma\, F_l \,.\, w_l\, \delta R$, where F_l is a generalized force, which as usual, is defined through $F_{lR} = -\partial E_l/\partial R$. Using our convention as to signs, we can identify $-\Sigma\, F_l \,.\, w_l\, \delta R$ as the work done *on* the system by the force. If we call it $-\delta W$, we have

$$\delta W = (\delta E)_{w_l = \text{const.}} = \sum_l F_l \,.\, w_l \delta R \qquad 3.65$$

Since we know w_l from the distribution, we can calculate δW for given F. We can make this more concrete by considering work against a constant pressure p. We can see that the r.h.s. of equation 3.65 defines the average force $\langle F \rangle$, or that

$$\delta E = -\langle F \rangle\, \delta R \qquad 3.66$$

But, by definition $p = \langle F \rangle / A$, where A = area of sub-system. Then

$$\delta E = -\, A p \delta R = -\, p \delta V \qquad 3.67$$

since $\delta V = A \delta R$. As pressure and volume are easily measurable, we now have a simple method of measuring δE or, strictly, part of δE.

Turning now to the second term on the r.h.s. of equation 3.64, we find that it is most simply evaluated by an indirect approach. First, we write the first law in variational form:

$$\delta E = \delta W + \delta Q \tag{3.68}$$

Because we have identified δW already, we see that the second term must be identical with δQ if the phenomenological and statistical approaches are to coincide. Then

$$\delta Q = \delta E - \delta W \tag{3.69}$$

and we can evaluate $\sum_l E_l \,.\, \delta w_l$ by evaluation of $\delta E - \delta W$ instead of directly.

Writing out δW in terms of the explicit form of w_l, we get

$$\delta Q = \delta E - \frac{\sum_l \exp(-\beta E_l)\, \Omega(E_l)\, \delta E_l}{Z} \tag{3.70}$$

To interpret the sum, we vary $\Sigma \exp(-\beta E_l)\, \Omega(E_l)$, getting

$$\delta\{\Sigma \exp(-\beta E_l)\, \Omega(E_l)\} = -\beta\, \Sigma \exp(-\beta E_l)\, \Omega(E_l)\, \delta E_l - \Sigma E_l \exp(-\beta E_l)\, \Omega(E_l)\, \delta\beta$$

where we have used the fact that $\Omega(E_l)$ = const. for constant R. Then

$$-\Sigma \exp(-\beta E_l)\, \Omega(E_l)\, \delta E_l = \delta \,.\, \frac{1}{\beta}\{\Sigma \exp(-\beta E_l)\, \Omega(E_l)\} + \frac{1}{\beta}\, \Sigma E_l \exp(-\beta E_l)\, \Omega(E_l)\, \delta\beta$$

On dividing by Z, we can identify the terms in equation 3.70 as

$$\delta Q = \delta E + \frac{1}{\beta}\frac{\delta Z}{Z} + \frac{\langle E\rangle}{\beta}\, \delta\beta \tag{3.71}$$

$$= \frac{1}{\beta}\, \delta\, (\langle E\rangle \beta + \ln Z) \tag{3.72}$$

where the second result follows from recalling that E and $\langle E\rangle$ mean the same quantity. Since $\beta = 1/kT$, we can rewrite equation 3.72 using the definition of S, equation 3.63, obtaining

$$\delta S = k\, \delta(\langle E\rangle\beta + \ln Z) \tag{3.73}$$

Equation 3.73 is an exact differential, so

$$S = k[\langle E\rangle\beta + \ln Z] + S_0 \tag{3.74}$$

where S_0 is an inessential constant which may be used to adjust the entropy to match any arbitrary scale. In statistical thermostatics, to save writing, we put it as zero. Equation 3.74 is the statistical definition of the entropy, but we shall soon find other expressions for it.

Returning now to the original Equation 3.64, we see that it can be written in the following ways:

$$\delta E = \delta W + \delta Q = -\langle F\rangle\, \delta R + \frac{1}{\beta}\, \delta(\langle E\rangle\beta + \ln Z) \tag{3.75}$$

where it must be emphasized that a particular process, the quasi-static process, is implied. The most important feature, is perhaps, that it furnishes a statistical interpretation of the concept of heat change, which in phenomenological thermodynamics is left in a nebulous state. The statistical result has no very obvious or intuitive physical significance, but the computational rule for obtaining it is completely definite.

A few words about the entropy. Does equation 3.74 agree with Nernst's theorem? As $T \to 0$, $Z \to \exp(-\beta E_0)\ \Omega(E_0)$, where $\Omega(E_0)$ is the number of phase-space sites available for the ground state or, in quantum physics, is the degeneracy of the ground state. The mean energy in the ground state is clearly E_0, so

$$\lim_{T \to 0} S \to k[\beta E_0 + \ln \Omega(E_0) - \beta E_0] = k \ln \Omega(E_0) = S_0 \qquad 3.76$$

Depending on the system, S_0 may or may not be zero, but the Nernst theorem is proved in general.

S can be expressed in another way, using the following property of the Gibbs distribution. We have repeatedly said that the most probable state is enormously more probable than any other state; thus, we can write an approximation for Z which is

$$Z \approx \exp(-\beta\langle E\rangle)\ .\ \Omega(\langle E\rangle) \qquad 3.77$$

This merely states that the only contribution which is important in Z is that due to the most probable energy, which in turn is the mean energy. Using equation 3.77, S is given by

$$S = k \ln \Omega(\langle E\rangle) \qquad 3.78$$

Equation 3.78 is the classical Boltzmann formula for S. It states that S is directly proportional to the logarithm of the number of states available, at the *mean energy*. In other words, S depends on the number of representative points on the surface of the mean-energy hypersphere. Equation 3.78 leads to the *principle of additivity* for entropy. If we evaluate equation 3.78 for m sub-systems taken together to form a system,

$$S_T = k \ln(\Omega) = k \ln \{\prod_m \Omega_m\} = k \sum_m \ln \Omega_m$$

$$= \Sigma S_m \qquad 3.79$$

Or, the entropy of the whole system is the sum of the entropies of the sub-systems. This additivity property is one of the major reasons for using entropy in thermodynamic calculations.

In thermodynamics, Helmholtz found it valuable to introduce a quantity called the 'Helmholtz free energy', defined by

$$F_H = \langle E\rangle - TS \qquad 3.80$$

Statistically, $F_H = -kT \ln Z$ 3.81

We shall find F_H useful in later calculations. Another free energy, the Gibbs free energy, is defined by

$$G = \langle E \rangle - TS + pV$$
$$= F_H + pV$$
$$= pV - kT \ln Z \qquad 3.82$$

We shall make no use of G, but both these quantities, which are often called 'thermodynamic potentials', are widely used in phenomenological thermodynamics. In view of the notation difficulty, one ought to remember that F_H is *not* a generalized force. Some properties of F_H and G are derived in Appendix 2.

To conclude the discussion of S, we shall put it in a form which applies when $\Omega(E_l)$ = unity, for all levels. Then

$$w_l = \frac{\exp(-\beta E_l)}{Z}$$

or $E_l = -\dfrac{1}{\beta}\ln(Zw_l)$

By definition, $\langle E \rangle = \sum_l w_l \,.\, E_l$

and $$\beta \langle E \rangle = \beta \sum_l w_l \,.\, E_l$$

$$= -\sum_l w_l \ln(Zw_l) = -\left\{\sum w_l \ln Z + \sum w_l \ \ln w_l\right\}$$

$$= -\left\{\ln Z + \sum w_l \ln w_l\right\}$$

since $\sum_l w_l = 1.0$, by definition. Then

$$S = -k\left[\sum_l w_l \,.\, \ln w_l\right] \qquad 3.83$$

$$= -k\,\langle \ln w_l \rangle \qquad 3.84$$

Apart from k, which is only a metric specifying the units, equation 3.83 arises in information theory as the definition of the information content of a source from which a finite set of messages is sent with probability w_l. Chapter 10 discusses the connection between information and statistical mechanics in more detail.

We have now reached a point from which it is possible to develop the necessary translations of macroscopic thermodynamics into statistical forms. A few particularly important relationships are given in the appendix; readers who are especially interested in these questions should refer to references 5, 10 and 11.

3.11 The properties of the partition function

From the last section it is obvious that the application of Gibbs' methods to particular systems comes down to the evaluation of the partition function Z in particular cases. We first look at the relationship between the partition function of an elementary sub-system and that of a system including many sub-systems. As an example, consider a free gas molecule; the partition function is

$$Z = \frac{1}{h^3} \int_0^\infty \exp(-\beta E) \frac{\partial \Gamma}{\partial E} \, dE \qquad 3.85$$

and we have shown that

$$\Gamma = V \cdot \frac{4\pi}{3} (\hat{p})^3, \quad \hat{p} = (2mE)^{\frac{1}{2}}$$

Substituting and integrating,

$$Z = \left(\frac{2\pi mkT}{h^2} \right)^{\frac{3}{2}} . V \qquad 3.86$$

when the value of β is inserted. In obtaining equation 3.56 we have really guessed that Z_N is directly proportional to Z^N, or $\ln(Z_N) \propto N \ln Z$. Is this correct and, if so, what is the factor of proportionality? Such factors dropped neatly out of equation 3.57 for the probability.

Using equation 3.86 to find S, we can write an explicit expression for the entropy of a free molecule. It is

$$S_1 = k \ln V + \frac{3k}{2} \ln T + \frac{3k}{2} \ln \left(\frac{2\pi mk}{h^2} \right) + \frac{3k}{2}$$

Since the entropy is an *additive* property, we ought to be able to write

$$S_N = NS_1 = Nk \ln V + \frac{3Nk}{2} + \frac{3Nk}{2} \left\{ \ln T + \ln \left(\frac{2\pi mk}{h^2} \right) \right\} \qquad 3.87$$

However, this expression cannot be correct because if one doubles N and doubles V, for example by removing a partition between two identical volumes of gas, one must double S_N, which is clearly not the case according to equation 3.87. If we put

$$Z = Z^N / N! \qquad 3.88$$

we must subtract $k \ln(N!) \approx kN(\ln N - 1)$ from equation 3.87 to get

$$S_N = Nk \ln \left(\frac{V}{N} \right) + \frac{5Nk}{2} + \frac{3Nk}{2} \left\{ \ln T + \ln \left(\frac{2\pi mk}{h^2} \right) \right\} \qquad 3.89$$

S_N given by equation 3.89 clearly does double when both N and V are doubled, and in the form

$$S_N = kN\left[\ln\left(\frac{V}{N}\right) + \frac{3\ln T}{2} + \sigma_0\right] \qquad 3.90$$

where σ_0 is a constant, is correct.

This discrepancy is called *'Gibbs' paradox'*. The difficulty is fundamental in classical statistics and arises from the problem of whether particular sorts of particle are to be considered as, in principle, *distinguishable* or *indistinguishable* from one another. We find it hard to remember nowadays that physicists in 1900 would hardly have recognized that a problem existed, because their thinking was so profoundly integrated with macroscopic concepts that they instinctively felt that molecules would, in some sense, be distinguishable. Quantum physics had to be well developed before the indistinguishability of microparticles was freely accepted.

Now, it is clear that molecules of the same gas are not, in fact, distinguishable from one another. No test will make a distinction, and labelling some particles by, say, ionization will change their mass, albeit by a trifling amount. Molecules of different ideal gases, say He and Ar, are equally clearly distinguishable, by their mass difference. Thus, the development of Gibbs statistics, which was largely based on Maxwell's ideas on ideal gases, is faulty in making an *implicit* assumption that the molecules are distinguishable. The difference comes out most clearly in equation 3.33, which includes the assumption. If the molecules are instead indistinguishable, then any arrangement which is merely a permutation of the N particles among themselves will be identical with all other such arrangements and we ought to divide equation 3.33 by $N!$, which is the number of such permutations. Clearly, this makes no difference for $N = 1$, so the one-molecule partition function is correct, although the derivation then seems completely artificial. Why has the method worked, with the exception of the final problem of the entropy of N particles? The answer really is that, by working with normalized probabilities, we have dropped inessential constants; for example, there is no term depending on $N!$ in the equations which fix the values of the Langrangian multipliers. The probabilities, then, are correct, but one must take care in the actual evaluation of properties of a specified mass of gas, by using the *correct* partition function for N molecules, viz.

$$Z_N = \frac{V^N}{N!}\left(\frac{2\pi mkT}{h^2}\right)^{3N/2} \qquad 3.91$$

In considering Fermi–Dirac and Bose–Einstein probability distributions we shall take indistinguishability into account from the start.

3.12 Reduction of the Gibbs distribution to the Maxwell–Boltzmann distribution

The Maxwellian distribution functions give, basically, the probabilities of finding a specified molecule in a specified range of energy or momentum or speed or velocity. The system is therefore a single molecule, in a heat reservoir. The Gibbs expression for the probability for this case is

$$\mathrm{d}w_{\mathrm{p}} = \frac{V}{Zh^3} \exp(-\beta E_{\mathrm{trans.}})\, \mathrm{d}p_1\, \mathrm{d}p_2\, \mathrm{d}p_3$$

With

$Z = (2\pi mkT/h^2)^{\frac{3}{2}}\ V$ from equation 3.86,

$$\mathrm{d}w_{\mathrm{p}} = \frac{\exp[-(\beta/2m)(p_1{}^2 + p_2{}^3 + p_3{}^2)]\ \mathrm{d}p_1\, \mathrm{d}p_2\, \mathrm{d}p_3}{(2\pi mkT)^{\frac{3}{2}}} \qquad 3.92$$

By converting to spherical coordinates and integrating over the angular coordinates, this becomes

$$\mathrm{d}w_{\mathrm{p}} = \frac{4\pi}{(2\pi mkT)^{\frac{3}{2}}} \exp\left(-\frac{\beta p^2}{2m}\right) p^2\, \mathrm{d}p \qquad 3.93$$

or, since $p = (2mE)^{\frac{1}{2}}$,

$$\mathrm{d}w_{\mathrm{E}} = \frac{2}{\pi^{\frac{1}{2}}} \cdot \frac{1}{(kT)^{\frac{3}{2}}} \exp(-\beta E)\, E^{\frac{1}{2}}\, \mathrm{d}E \qquad 3.94$$

Now, these are the probabilities of finding one molecule in a specific state. Thus, the expected number of molecules in this state when N are present is $N \times \mathrm{d}w$.

For energy,

$$\mathrm{d}N_{\mathrm{E}} = \frac{2N}{\pi^{\frac{1}{2}}} \cdot \frac{1}{(kT)^{\frac{3}{2}}} \exp(-\beta E)\, E^{\frac{1}{2}}\, \mathrm{d}E \qquad 3.95$$

This is exactly what Maxwell deduced from the considerations of Chapter 2, and is identical with equation 2.23. It is hardly necessary to reproduce the remaining expressions, and this is left as an exercise for the reader.

3.13 The grand canonical distribution and the constant μ

We were forced to defer the further discussion of the grand canonical distribution until we had achieved statistical definitions of entropy etc. We reopen the discussion by identifying the constant μ, which is called the '*chemical potential*.' The technique used for evaluating μ starts with a repetition of the reasoning by which we showed that the interaction of two sub-systems had to be described by a uniform value of β. We need not write down the details as it can easily be verified that equations equivalent to equations 3.48 and 3.49 which use the grand canonical expressions for W_1 and W_2 must imply not only that $\theta_1 = \theta$ but also that $\mu_1 = \mu_2 = \mu$.

It will probably be helpful to summarize these conditions by saying that they mean that, when the sub-systems exchange both energy and particles, the gain and loss of energy must balance and the number of particles gained and lost must balance. In turn, this implies that the mean energies of the particles involved in the exchange are the same.

Turning now to the evaluation of μ, which was originally defined by equation 3.27, we see that, if we use equation 3.78 to eliminate $\ln(\Omega)$,

$$\mu = -T(\partial S/\partial N)_{N=N_0} \qquad 3.96$$

This result is correct, but it is informative to derive it in a more general manner, which also provides alternative expressions for μ.

Let a quasi-static sub-system comprise N particles. Then, per particle, the statistical equivalent of the first law of thermodynamics is

$$\delta\left(\frac{E}{N}\right) = -p\,\delta\left(\frac{V}{N}\right) + \frac{1}{k\beta}\,\delta\left(\frac{S}{N}\right) \qquad 3.97$$

where we have used equations 3.67, 3.72 and 3.73.

In a change which is due to a variation in N, carrying out the variation and collecting terms,

$$dE = \frac{dS}{k\beta} - p\,dV + \left(\frac{E + pV - S/k\beta}{N}\right) dN \qquad 3.98$$

The third term is new and arises from the assumption that N changes – which, of course, is what was assumed in the grand canonical distribution.

Assume that

$$\mu = \left(\frac{E + pV - S/k\beta}{N}\right) \qquad 3.99$$

then, if S and V are constant, equation 3.98 gives

$$\mu = (\partial E/\partial N)_{S,\,V\ \mathrm{const.}} \qquad 3.100$$

If E and V are constant,

$$\mu = -\frac{1}{k\beta}\left(\frac{\partial S}{\partial N}\right)_{E,\,V\ \mathrm{const.}} \qquad 3.101$$

Then, equation 3.101 confirms the correctness of equation 3.90 and clarifies the meaning of taking the derivative at $N = N_0$. Naturally, this means that the assumption equation 3.99 is confirmed. We can therefore rewrite equation 3.98 as

$$dE = \frac{dS}{k\beta} - p\,dV + \mu dN \qquad 3.102$$

From equation 3.80 we find that $dF_H = dE - d(TS)$, which alters equation 3.102 to

$$dF_H = -S dT - p dV + \mu dN$$

giving $\mu = (\partial F_H/\partial N)_{T,\,V\ \mathrm{const.}}$ 3.103

Equation 3.103 is useful because, especially in chemical thermodynamics, the experimental behaviour of F_H is extensively investigated.

Exercise

Show that F_H and μ for an ideal gas are given by

$$F_H = -NkT \ln \left[V . \left(\frac{2\pi mkT}{h^2} \right)^{\frac{3}{2}} \right] + kT \ln N!$$

$$\approx -NkT \ln \left[\frac{eV}{N} \left(\frac{2\pi mkT}{h^2} \right)^{\frac{3}{2}} \right] \qquad 3.104$$

$$\mu = kT \ln \left[\frac{N}{V} \left(\frac{h^2}{2\pi mkT} \right)^{\frac{3}{2}} \cdot \frac{1}{(kT)^{\frac{3}{2}}} \right] \qquad 3.105$$

Hint:
It is easier to start from equation 3.81 than from 3.80. Calculation shows that the bracket term is $\ll 1$, so that μ is large and *negative* in this case.

3.14 The equipartion of energy

The reader will have noticed that our examples have dealt only with translational energy. It is quite easy to extend Gibbs' methods to include many other cases. Examples are the mechanics of diatomic molecules, simple crystals and (much harder) phase changes. These questions are discussed in detail, in particular, in those books with a chemical emphasis (references 6–10); however, the classical equipartition theorem is so important that it merits discussion here.

The theorem arises from the observation that in the simple translational case a mean energy of $kT/2$ is associated with each of the three degrees of freedom of a particle. Can this result be generalized, in classical theory, to cover more complicated situations? The restriction to classical theory is obviously necessary since quantum systems of even the simplest type do not obey the theorem. However, classically it is possible to show that the mean value of each quadratic term in the energy (Hamiltonian) is $kT/2$, and this constitutes the equipartition theorem.

To prove this, we consider a system in which there are more than three degrees of freedom and the energy is no longer that of pure translation. An example would be a diatomic molecule which has three degrees of freedom and which can also rotate about an axis joining the centres of the molecules, therefore having five degrees of freedom in all. As usual, the system is described by f generalized co-ordinates and f momenta.

It is now assumed (which is normally found to be the case) that the total energy is

1. quadratic in either the momenta or the co-ordinates
2. of such a form that it can be split into two additive terms, one containing the momentum (or coordinate) which we have decided to single out for attention and the other containing all the remaining contributions.

To continue the proof, we shall consider the j th component of momentum. Then, we assume that

$$E_{\mathrm{T}} = E'(p_{\mathrm{j}}) + E''(q_{\mathrm{i}} \ldots q_{\mathrm{f}}, p_{\mathrm{i}} \ldots \not{p}_{\mathrm{j}}\, p_{\mathrm{f}})$$

Here, the notation stresses that E'' does *not* depend on p_{j}. In view of the restriction to quadratic terms, $E'(p_{\mathrm{j}}) = \alpha p_{\mathrm{j}}^2$, where α is a constant. Then, by definition, the Boltzmann theorem gives the mean value of $E'(p_{\mathrm{j}})$ as

$$\langle E_{\mathrm{j}}{}^1 \rangle = \frac{\int E'_{\mathrm{j}} \exp(-\beta E_{\mathrm{T}})\, (\partial\Gamma/\partial E)\, \mathrm{d}E}{\int \exp(-\beta E_{\mathrm{T}})\, (\partial\Gamma/\partial E)\, .\, \mathrm{d}E}$$

$$= \frac{\int E'_{\mathrm{j}} \exp(-\beta E'_{\mathrm{j}})\, \mathrm{d}p_{\mathrm{j}} \int \exp(-\beta E'')\, (\partial\Gamma''/\partial E)\, \mathrm{d}E}{\int \exp(-\beta E'_{\mathrm{j}})\, \mathrm{d}p_{\mathrm{j}} \int \exp(-\beta E'')\, (\partial\Gamma''/\partial E)\, \mathrm{d}E}$$

where Γ'' is the number of representative points after omitting the p_{j} coordinate.

$$\therefore\ \langle E'_{\mathrm{j}} \rangle = \frac{\int E'_{\mathrm{j}} \exp(-\beta E'_{\mathrm{j}})\, \mathrm{d}p_{\mathrm{j}}}{\int \exp(-\beta E'_{\mathrm{j}})\, \mathrm{d}p_{\mathrm{j}}} = -\frac{\partial}{\partial\beta} \ln\left(\int \exp(-\beta E'_{\mathrm{j}})\, \mathrm{d}p_{\mathrm{j}}\right)$$

the last result being suggested by equation 3.20. Now,

$$\int \exp(-\beta E'_{\mathrm{j}})\, \mathrm{d}p_{\mathrm{j}} = \int_{-\infty}^{+\infty} \exp(-\beta\alpha p_{\mathrm{j}}^2)\, \mathrm{d}p_{\mathrm{j}} = (\pi/\alpha\beta)^{\frac{1}{2}}$$

and $\qquad \ln\left[\int\right] = \tfrac{1}{2}\left[\ln(\pi/\alpha) - \ln\beta\right]$

therefore $\qquad \langle E'_{\mathrm{j}} \rangle = -\dfrac{\partial}{\partial\beta} \ln\left[\int\right] = \dfrac{1}{2\beta}$

$$= \frac{kT}{2}$$

This was the result we wished to prove.

The equipartition theorem is extremely useful in view of the wide range of circumstances in which it can be applied. In particular, it can be applied to the macroscopic electric and magnetic energies of electromagnetic systems and is frequently used for the deduction of fluctuation values in electromagnetic circuits.

Conclusion

We have now discussed the major features of the Gibbs method. The books listed as references contain the details of applications to many more examples than we can discuss here. However, our main objective is to isolate the method from the details of its application and we must now show the modifications imposed by the study of quantal systems, which are the subject of the next two chapters.

4

Quantum Statistics: the Fermi–Dirac Distribution

In the last chapter we have already discussed two important differences between classical statistics and quantum statistics. These are

1. The fact that microparticles of the same species are indistinguishable from one another; thus, all Ar molecules are identical, all electrons are identical, etc.

2. In quantum theory the energy levels are discrete, although for large energies the spacing between levels may be negligibly small.

To these we must add two other important conditions which lead to probability distributions with widely different properties. The first condition, leading to Fermi-Dirac statistics, follows.

Quantum theory shows that, when a particle has a total spin angular momentum which is given by $\hbar/2$, $3\hbar/2$, etc., there is a fundamental restriction on the total wave function. This is that the wave function must be antisymmetrical, i.e. that, if we exchange two particles with one another, the wave function changes sign. In this case, there can never be a state of the 'gas' of particles in which there are two (or more) particles in the same single-particle state. Alternatively, we may say that there can be at most two particles with identical quantum numbers *l, m, n* and equal positive and negative spins. This principle is called, after its discoverer, 'Pauli's exclusion principle'. Particles which obey the principle are statistically distributed according to the Fermi–Dirac distribution and, nowadays, are often given the general name 'fermions'. By far the most common example of a fermion is the electron, either free or in a metal or semiconductor.

If we look back to the occupation numbers introduced when we discussed the method of the most probable distribution in the last chapter (section 3.7), we have now said that, for fermions, the occupation numbers can have only the values 0 or 1.

4.1 The Fermi distribution – simple derivation

This derivation is so simply done directly that it is well worth considering it before attempting more formal derivations. Consider, for concreteness, a small rectangular block of metal, volume V, and assume that a very large number N of (nearly) free electrons are in a state of molecular chaos in the volume, colliding with the lattice ions, with the barriers at the walls of the parallelepiped and with one another. Let n_s electrons have energy in dE_s at E_s and let the number of states available for occupation be g_s. For fermions, we must have $n_s \leqslant g_s$ since, at most, one particle can be in each state. Then, in this s th level, n_s states are filled while $g_s - n_s$ are empty. The number of ways in which this can happen with identical particles is the number of combinations of g_s things taken n_s at a time, viz.

$$W_s = \frac{g_s!}{n_s!\,(g_s - n_s)!} \tag{4.1}$$

A similar result is obtained for every other energy range. Each combination may occur together with any other combination for a different energy level. For example, we can write

$$W_s \,.\, W_k = \frac{g_s!}{n_s!\,(g_s - n_s)!} \cdot \frac{g_k!}{n_k!\,(g_k - n_k)!}$$

or, in general,

$$W = \prod_s \frac{g_s!}{n_s!\,(g_s - n_s)!} \tag{4.2}$$

We can now find the maximum value of W or the most probable arrangement by taking ln W and expansion using Stirling's approximation and the use of Lagrangian multipliers $-\alpha$ and $-\beta$ (which are not necessarily the same as those used in Chapter 3 and whose signs are chosen with hindsight). Then

$$\ln W = \Sigma \{\ln g_s! \; - \ln n_s! - \ln (g_s - n_s)!\}$$
$$\approx \Sigma [g_s \ln g_s - n_s \ln n_s - (g_s - n_s) \ln (g_s - n_s)]$$

The subsidiary conditions are

$$\sum_s n_s = N \tag{4.3}$$

$$\sum_s E_s n_s = E \tag{4.4}$$

Varying ln W with respect to the n_s, we find

$$\Sigma \{-\delta n_s \ln n_s + \delta n_s \ln(g_s - n_s)\} = 0$$

or
$$n_s = \frac{g_s}{1 + \exp(\alpha + \beta E_s)} \tag{4.5}$$

Now, we have already said that $n_s \leqslant g_s$. Consider for a moment the situation when E_s is very large (compared with kT, for example); the number of states g_s will then be very large according to the considerations of Chapter 3. It is not unreasonable to think that n_s ought to be much less than g_s in this case. This happens if $\exp(a + \beta E_s) \gg 1$, since that gives

$$n_s = g_s \,.\, \exp - (\alpha + \beta E_s) \tag{4.6}$$

Equation 4.6 is formally identical with Boltzmann's theorem if $\beta = 1/kT$. According to the correspondence principle, equation 4.5 should tend to the classical result when E_s is large, so this is a strong argument for believing that $\beta = 1/kT$. From the form of equation 4.5 it is obvious that α is a pure number; therefore we can rewrite $\alpha = E_F/kT$, where E_F is another energy constant which remains to be determined.

What about the sign of α or E_F? Let $T \to 0$, for arbitrary E_s. Then $\exp(E_F + \beta E_s) \to \infty$, if E_F has a positive sign. Equation 4.5 then states that n_s is always zero, which is nonsense. Therefore, E_F has a negative sign and we have to consider $\exp[E_s - E_F)/kT]$. If $E_F > E_s$, this is zero. When $E_F = E_s$ a discontinuity occurs, and when $E_s > E_F$ the exponential is again infinite. Thus, n_s changes abruptly from $n_s = g_s$ to $n_s = 0$ at $E_s = E_F$, see Fig. 4.1

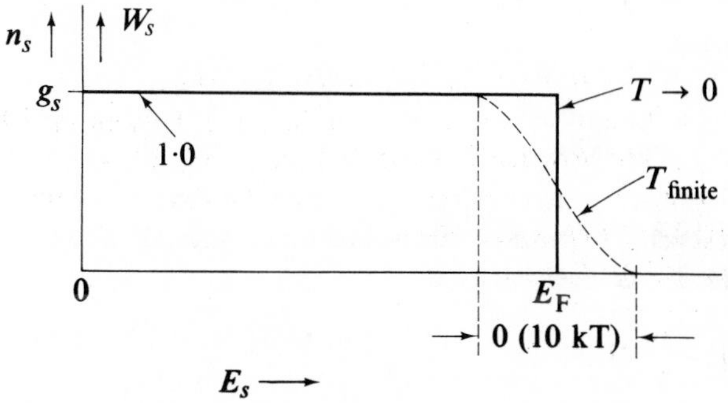

Fig. 4.1 The Fermi-Dirac probability function for absolute zero and for a finite temperature.

Now, consider that kT is small, equal to a number δ. When $E_s - E_F > 5\delta$, the exponential is much greater than unity and n_s is $\ll g_s$. Conversely, when $E_F - E_s > 5\delta$, the exponential is very small and $n_s \approx g_s$. Thus, n_s drops from g_s to zero over a range of energy of order 10 kT, centred on E_F. The *probability* of finding a state at E_s filled, which is n_s/g_s, similarly drops from unity for $E_s \ll E_F$, to zero for $E_s \gg E_F$. It is *defined* as equal to $\frac{1}{2}$ when $T \to 0$ and $E_s = E_F$, in accordance with the usual mathematical definition of the measure of a jump from 1.0 to 0.

To summarize the results so far, we have strong reasons for believing that $\beta = 1/kT$ and is identical with the β of the Maxwell–Boltzmann distribution. We have introduced a normalizing energy E_F whose value we do not yet know how to calculate, and we have shown that the signs in the exponential must be such that we can write the probability of finding a filled state at E_s as

$$f(E_s) = \frac{1}{1 + \exp[(E_s - E_F)/kT]} \qquad 4.7$$

This is called the 'Fermi–Dirac probability distribution function', or 'F–D function' for short.

Returning to equations 4.3 and 4.4, we have

$$\sum g_s f(E_s) = \sum \frac{g_s}{1 + \exp[(E_s - E_F)/kT]} = N \qquad 4.8$$

and $$g_s E_s f(E_s) = \sum \frac{g_s E_s}{1 + \exp[(E_s - E_F)/kT]} = E \qquad 4.9$$

and we can use these to determine E_F and the mean energy $\langle E \rangle = E/N$.

Let us now calculate g_s as a function of E, by a direct method which is similar to the methods of Chapter 3, but not quite identical. We think of our metal block, volume V, sides length a_1, a_2, a_3. The energy eigenvalues are then

$$E_{n,\,m,\,l} = \frac{\pi^2 \hbar^2}{2m} \left[\frac{n^2}{a_1{}^2} + \frac{m^2}{a_2{}^2} + \frac{l^2}{a_3{}^2} \right]$$

which is conveniently rewritten

$$\frac{\hbar^2}{2m} \left[k_1{}^2 + k_2{}^2 + k_3{}^2 \right] \qquad 4.10$$

where the k's are the wave numbers or, to engineers, propagation constants, of the probability waves and are restricted to *positive* values. In a small range, dn at n, dm at m, dl at l, there are dn.dm.dl eigenvalues. But

$$\mathrm{d}n\,.\,\mathrm{d}m\,.\,\mathrm{d}l = \frac{a_1}{\pi}\,\mathrm{d}k_1 \,.\, \frac{a_2}{\pi}\,\mathrm{d}k_2 \,.\, \frac{a_3}{\pi}\,\mathrm{d}k_3$$

$$= \frac{V}{\pi}\,.\,\mathrm{d}k_1\,\mathrm{d}k_2\,\mathrm{d}k_3 \qquad 4.11$$

Now, imagine k_1, k_2, k_3 plotted in three-dimensional k space. Any eigenvalue must be represented by a point on the surface of a sphere radius $K = \sqrt{(k_1{}^2 + k_2{}^2 + k_3{}^2)}$. The volume element of this sphere is $4\pi K^2\,\mathrm{d}K$; but, if we are restricted to positive values of k_1, k_2, k_3, only one octant of the sphere is allowed, since all the other octants include *negative* k values. Thus, in our problem, the relevant K-space volume is

$$\frac{\pi}{2} K^2\,\mathrm{d}K \qquad 4.12$$

But, from equation 4.10, $K^2\,\mathrm{d}K = 2^{1/2} m^{3/2} E^{1/2}\,\mathrm{d}E/h^3$ and the density of states allowed in dE at E, dividing equation 4.11 by V, is

$$\frac{m^{3/2} E^{1/2}\,\mathrm{d}E}{2^{1/2} \hbar^3 \pi^2} \qquad 4.13$$

For *electrons,* spin values of $\pm \frac{1}{2}$ are allowed and so we must double this to get

$$g_s(e_-) = \frac{2^{1/2} m^{3/2} E^{1/2}\,\mathrm{d}E}{\hbar^3 \pi^2} \qquad 4.14$$

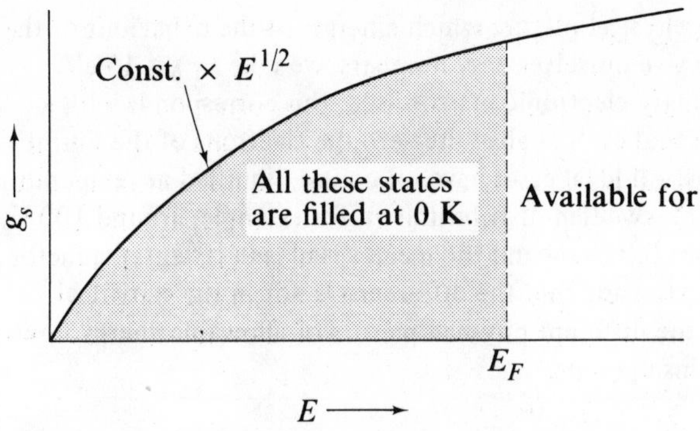

Fig. 4.2 The number of states available, as a function of energy at absolute zero.

which varies as $E^{1/2}$, as indicated in Fig. 4.2. The number of available states is then a monotone increasing function of E.

From the earlier discussion, it is clear that all those states with energy below E_F are occupied at $T = 0$ K. At higher temperatures, some electrons will be found in higher energy states and some of the states with energies $< E_F$ will be found empty.

It is now a simple matter for us to compute a value for E_F which is correct for $T \to 0$ and, as we shall show later, is the first term in a series expansion for higher temperatures. If E_{F0} = value of E_F at 0 K and there are n_e electrons per unit volume (n_e is therefore the electron density), we have, using equation 4.8,

$$n_e = \frac{(2m^3)^{1/2}}{\hbar^3\pi^2} \int_0^\infty \frac{E^{1/2}\,dE}{1 + \exp[(E - E_F)/kT]}$$

$$\to \frac{(2m^3)^{1/2}}{\hbar^3\pi^2} \int_0^{E_{F0}} E^{1/2}\,dE, \text{ as } T \to 0$$

since the probability distribution is just the step function in Fig. 4.1.

$$\text{Then} \quad E_{F0} = \frac{\hbar^2}{2m}(3\pi^2 n_e)^{2/3} \tag{4.15}$$

which depends only on universal constants and the two-thirds power of the electron density. We remark that for most metals, E_{F0} is of the order 10 electron volts, whose equivalent temperature is over 10^5 K* so that E_{F0} is very much greater than kT for all laboratory conditions.

The reader may make an exactly similar calculation to obtain $\langle E \rangle$ from equation 4.9. The result is

$$\frac{\langle E_0 \rangle}{n_e} = \frac{3\hbar^2}{10m}(3\pi^2 n_e)^{2/3} = \frac{3E_{F0}}{5} \tag{4.16}$$

or, the mean energy per electron is 0.6 times the Fermi energy.

*The temperature equivalent of E_{F0} is often called the 'Fermi temperature', T_F. For copper, T_F is almost exactly 8×10^4 K.

Let us briefly discuss the physical picture which emerges of the behaviour of the electrons in the metal. If, to give ourselves easy numbers, we take $E_{F0} = 15$ eV, then $\langle E_0 \rangle = 9$ eV. If the ordinary electronic mass is used, this corresponds with a velocity of 1.8×10^6 m/s, so that even at absolute zero the electrons of the Fermi sea move with high velocities. All ideal gases have, of course, liquified at temperatures above a few kelvin, but in a Maxwellian distribution at, for example, around 100 K, the mean energy is only about 0.015 eV and the mean speed is $0(10^4$ m/s), a factor of 100 or so less. It is worth stressing that the difference is not in the statistical reasoning employed, it is in the different physical picture of allowable energy levels which quantum theory enjoins upon us.

4.2 The Fermi distribution – derivation from Gibbs' grand canonical ensemble*

In this method we start off from the fundamental proposition that, in quantum theory, all we can say *ab initio* about the distribution of fermions is that there are

n_1 particles in the state with energy E_1,
n_2 particles in the state with energy E_2,
n_3 particles in the state with energy E_3,
n_r particles in the state with energy E_r,

all particles being identical.

Also, we know that

$$\sum_r n_r = N \qquad 4.17$$

and the total energy is $E_T = (n_1 E_1 + n_2 E_2 + \ldots + n_r E_r + \ldots)$

$$= \sum_r n_r E_r \qquad 4.18$$

From Gibbs' grand canonical distribution (section 3.6) we can find the expression for the mean number of particles which have a particular value of the energy. It is

$$\langle n \rangle = kT \frac{\partial}{\partial \mu} \ln \sum_r \sum_{n_r} \exp\left(\frac{\mu n - E_r}{kT}\right) \Omega(E_r, n_r)$$

However, in this case by writing out the terms and, in view of equation 4.18, we can simplify this to

$$\langle n \rangle = kT \frac{\partial}{\partial \mu} \ln \sum_{n_r} \exp\left(\frac{\mu - E_r}{kT}\right)^{n_r} \Omega(n_r) \qquad 4.19$$

Next, we recall the meaning of $\Omega(n_r)$: it is the density of states per unit energy

*Many readers may wish to omit this section on a first reading.

range divided by, for a single fermion, h^3. But the volume of phase space corresponding to a single quantum state is also h^3, so $\Omega = 1.0$. Thus

$$\langle n\rangle = kT\frac{\partial}{\partial\mu}\ln\left[\exp\left(\frac{\mu - E_r}{kT}\right)^0 + \exp\left(\frac{\mu - E_r}{kT}\right)^1\right]$$

since, for fermions r can have only the values 0 or 1. Therefore

$$\langle n\rangle = kT\frac{\partial}{\partial\mu}\ln\left[1 + \exp\left(\frac{\mu - E_r}{kT}\right)\right]$$

$$= \exp\left(\frac{E_r - \mu}{kT} + 1\right)^{-1} \qquad 4.20$$

$$= \frac{1}{\exp[(E_r - \mu/kT) + 1]} \qquad 4.21$$

Alternatively, if we assume, as is always the case, that the energy spectrum is continuous for high energies, we can write dn, the number of particles in dE at E, in the form

$$\mathrm{d}n = \frac{1}{1 + \exp[(E_r - \mu)/kT)]} \cdot \frac{\partial\gamma}{h^3} \qquad 4.22$$

Thus, if we indentify μ with E_F, the derivations lead to identical results.

4.3 Mathematical treatment of the Fermi function at low and high temperatures

Before moving to applications of the Fermi–Dirac distribution, it will be useful to develop some approximations for the Fermi function $f = [\exp(E - E_F)/kT + 1]^{-1}$. For example, if we wish to calculate the mean energy for $kT \neq 0$, we need to evaluate $\int_0^\infty E^{3/2} f(E)\,\mathrm{d}E$; therefore, consider integrals of the form $\int_0^\infty E^n f(E)\,\mathrm{d}E$.

Integrating by parts, we get

$$I_n = \left[f \cdot \frac{E^{n+1}}{n+1}\right]_0^\infty - \frac{1}{n+1}\int_0^\infty E^{n+1}\frac{\partial f}{\partial E}\cdot \mathrm{d}E$$

$$= -\frac{1}{n+1}\int_0^\infty E^{n+1}\frac{\partial f}{\partial E}\cdot \mathrm{d}E$$

since $E^{n+1} \to 0$ as $E \to 0$, and $f(E) \to 0$ as $E \to \infty$. The form of $\partial f/\partial E$ is indicated on Fig. 4.3. Clearly, $\partial f/\partial E$ is a Dirac delta function when $T = 0$ and can be represented

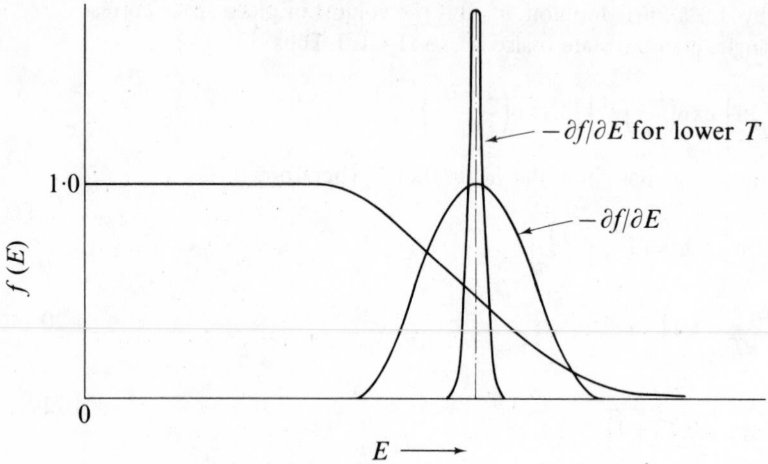

Fig. 4.3 Illustrating the method of approximation for the Fermi function at temperatures above zero.

by any of the functions which themselves define the delta function by a limiting process. Let $(E - E_F)/kT = u$; then

$$I_n = -\frac{1}{n+1}\int_{E_F/kT}^{\infty} (E_F + kTu)^{n+1}\frac{\partial f}{\partial u}.\,du$$

$$= -\frac{E_F{}^{n+1}}{n+1}\int_{-\infty}^{+\infty}\left(1 + \frac{kTu}{E_F}\right)^{n+1}\frac{\partial f}{\partial u}.\,du$$

since, from Fig. 4.3, the lower limit can be extended to $-\infty$ without altering the value of the integral. We recall that $f(E)$ changes over a range of about 10 kT, so the maximum range of u is $\pm$ 5 kT. Thus, kTu/E_F is $0\{5(kT)^2/E_F\}$, which is $\ll 1$ for any reasonable laboratory temperatures. We can therefore expand using the binomial theorem to get

$$I_n = -\frac{E_F{}^{n+1}}{n+1}\int_{-\infty}^{+\infty}\left[1 + (n+1)\frac{kTu}{E_F} + \frac{(n+1)n}{2}\left(\frac{kTu}{E_F}\right)^2 + \ldots\right]\frac{\partial f}{\partial u}.\,du$$

From the symmetrical nature of $\partial f/\partial E$, we see that

$$\int_{-\infty}^{+\infty} u\frac{\partial f}{\partial u}\,du = 0$$

$$\int_{-\infty}^{+\infty} u^2\frac{\partial f}{\partial u}.\,du = 2\int_{0}^{\infty} u^2\frac{\partial f}{\partial u}.\,du$$

$$\int_{-\infty}^{+\infty} u^3\frac{\partial f}{\partial u}.\,du = 0, \quad \text{etc.}$$

$$\therefore I_n = \frac{-E_F{}^{n+1}}{n+1}\left[\int_{-\infty}^{+\infty}\frac{\partial f}{\partial u}\,.\,du + (n+1)\,n\left(\frac{kT}{E_F}\right)^2\int_0^{\infty}u^2\frac{\partial f}{\partial u}\,.\,du + \ldots\right]$$

$$\approx \frac{E_F{}^{n+1}}{n+1}\left[1 + \frac{(n+1)n}{6}\pi^2\left(\frac{kT}{E_F}\right)^2\right] \qquad 4.23$$

since the integral of the delta function is -1 and the second integral can be shown to be $-\pi^2/6$ by expansion of $f(u)$ as a power series. We have just seen that there is no term in $(kT/E_F)^3$ and the next term is in $(kT/E_F)^4$, so the approximation is good especially for n fairly small. It is excellent when $T \ll T_F$.

By reference to section 4.1, we see that n_e is given by $\dfrac{(2m^3)^{1/2}}{\hbar^3\pi^2}\,.\,I_{1/2}$ or

$$n_e = \frac{(2m^3)^{1/2}}{\hbar^3\pi^2}\,.\,\frac{2E_F{}^{3/2}}{3}\left[1 + \frac{\pi^2}{8}\left(\frac{kT}{E_F}\right)^2 + 0(T^4)\right] \qquad 4.24$$

Similarly, $\langle E\rangle$ depends on $I_{3/2}$ and is given by

$$\langle E\rangle = \frac{(2m^3)^{1/2}}{\hbar^3\pi^2}\,.\,\frac{2E_F{}^{5/2}}{5}\left[1 + \frac{5\pi^2}{8}\left(\frac{kT}{E_F}\right)^2 + 0(T^4)\right] \qquad 4.25$$

The general procedure for handling these is that n_e is assumed known; we can, for example, calculate the density of atoms for a particular metal and, if one electron is freed per atom, the number density of electrons will equal the density of atoms. Equation 4.24 can then be solved for E_F by successive approximation, starting from $E_F \approx E_{F0}$ in the correction term. When a sufficiently accurate value of E_F has been found, it is used in equation 4.25 to obtain a value of $\langle E\rangle$ with similar accuracy. Naturally, the approximations given here, while very useful for accurate numerical computation, are not very valuable for further analysis, except in so far as they indicate the order of magnitude of the neglected terms. However, we can easily get some less accurate but analytically more useful approximations as follows.

Using equation 4.15 for n_e in terms of E_{F0}, we get

$$E_{F_0}^{3/2} \approx E_F{}^{3/2}\left[1 + \frac{\pi^2}{8}\left(\frac{kT}{E_{F0}}\right)^2\right]$$

$$\text{or} \quad E_F \approx E_{F0}\left[1 + \frac{\pi^2}{8}\left(\frac{kT}{E_{F0}}\right)^2\right]^{-2/3}$$

and, using the binomial expansion, this becomes

$$E_F \approx E_{F0}\left[1 - \frac{\pi^2}{12}\left(\frac{kT}{E_{F0}}\right)^2\right] \qquad 4.26$$

Notice that the Fermi level is *lowered* by increasing temperature.

It can be shown, by the same steps, that

$$\frac{\langle E\rangle}{n_e} \approx \frac{3}{5}E_{F0}\left[1 + \frac{5\pi^2}{12}\left(\frac{kT}{E_{F0}}\right)^2\right] \qquad 4.27$$

which formula is often quoted. For example, the electronic specific heat at constant volume, $C_V \div n_e$, is

$$\frac{d}{dT}\left[\frac{\langle E\rangle}{n_e}\right] = \frac{\pi^2 k^2 T}{2E_{F0}} \qquad 4.28$$

The classical result obtained by the use of the Maxwell–Boltzmann distribution is $3k/2$, i.e. independent of T. The linear temperature variation of T is experimentally verified, at temperatures of a few K where the lattice contribution to the specific heat does not mask the electronic specific heat. This was one of the early successes of Fermi–Dirac treatments. Numerically, equation 4.28 is not very near the experimental values for all metals, but in a more sophisticated theory allowance would be made for the replacement of the electronic mass by the effective mass m* and this improves the agreement considerably. The question is important in metal physics since it is an important technique for obtaining information about the shape of the Fermi surface.

So far, we have considered the Fermi distribution at low temperatures. For completeness we ought to look at the high-temperature form and also consider the conditions which govern the transition. Two different physical models are important. In the first, the electron gas is at a temperature which is large compared with the Fermi temperature T_F. It will be remembered that T_F is $0(10^5\ \text{K})$, so the temperatures are very high indeed. However, temperatures of $0(10^7\ \text{K})$ are recorded in plasma-fusion experiments, and very high temperatures are encountered in astrophysics; thus, situations in which $T/T_F \gg 1$ are encountered in special fields.

The second model, which is much more common, is used to obtain approximate expressions for the Fermi–Dirac distribution 'tail', that is those states which have energies a few times kT greater than E_F. The most obvious example is that of thermionic emission, in which the electrons liberated from the hot metal must have had energies greater than E_F + thermionic work function ϕ inside the hot metal. Since ϕ ranges up to about half the value of E_F, the energies in question are about $2E_{F0}$. In this case, the value of T/T_F is still rather small, for example at the melting point of tungsten it is about 0.04.

To examine these questions, we rewrite the Fermi function as

$$g(s)\,.\,f(E_s) = \frac{g(s)}{1 + \exp[(E_s - E_F)/kT]} = \frac{g(s)\exp[(E_F - E_s)/kT]}{\exp[(E_F - E_s)/kT] + 1} \qquad 4.29$$

Case 1: $kT \gg |E_{F0}.|$

We know that E_F decreases with increasing temperature, from equation 4.26. From that equation, used well beyond its limits of accuracy, $E_F \approx 0$ when $(kT/E_{F0}) \approx 1.1$, and becomes negative for still higher temperatures. Let us assume that E_F does, in fact, become negative; the r.h.s. of equation 4.29 then reduces to $g(s)\exp -[(E_F + E_s/kT)]$ whatever the value of E_s, as long as E_F/kT is not too small. We can now find

E_F in a manner similar to that employed in finding E_{F0} in equation 4.15. The details are (E_F is negative)

$$n_s = \frac{(2m^3)^{1/2}}{\hbar^3\pi^2}\int_0^\infty \exp\left(-\frac{E_F}{kT}\right).E_s^{1/2}\exp\left(-\frac{E_s}{kT}\right)dE_s$$

$$= \frac{(2m^3)^{1/2}}{\hbar^3\pi^2}\exp\left(-\frac{E_F}{kT}\right)\int_0^\infty E_s^{1/2}\exp\left(-\frac{E_s}{kT}\right)dE_s$$

The infinite upper limit contributes almost nothing to the integral after $E_s >$ several E_{F0}.

$$n_s = \frac{(2m^3)^{1/2}}{\hbar^3\pi^2}\exp\left(-\frac{E_F}{kT}\right).2(kT)^{3/2}\int_0^\infty u^2\exp(-u^2)\,du$$

$$= \frac{(2m^3)^{1/2}}{\hbar^3\pi^2}\exp\left(-\frac{E_F}{kT}\right).2(kT)^{3/2}.\frac{\sqrt{\pi}}{4}$$

or
$$\frac{(2m^3)^{1/2}}{\hbar^3\pi^2}.\frac{2E_{F0}^{3/2}}{3} = \frac{(2m^3)^{1/2}}{\hbar^3\pi^2}\exp\left(-\frac{E_F}{kT}\right).2(kT)^{3/2}\frac{\sqrt{\pi}}{4}$$

using equation 4.15.

$$\therefore \exp\left(\frac{E_F}{kT}\right) = \ln\frac{3\sqrt{\pi}}{4}\left(\frac{kT}{E_{F_0}}\right)^{3/2}$$

or
$$\frac{E_F}{kT} = \ln\frac{3\sqrt{\pi}}{4} + \frac{3}{2}\ln\left(\frac{kT}{E_{F0}}\right) \tag{4.30}$$

For example, $E_F \approx -13.5\, E_{F0}$ when $kT/E_{F0} = 5$.

We can now rewrite the high-temperature form of equation 4.29 as

$$g_1(s) = \exp(-E_s/kT) \tag{4.31}$$

with
$$g_1(s) = g(s)\exp(-|E_F|/kT) \ll g(s) \tag{4.32}$$

The situation is as sketched in Fig. 4.4(b), where $g_1(s)$ is a few percent of $g(s)$. It is a Maxwell–Boltzmann distribution with a very low broad peak.

Case 2: $kT \ll E_{F0}, E_s > E_{F0}$

Here, wherever E_s exceeds E_F by more than about 5 kT, we can once more put

$$g(s)f(E_s) \approx g(s)\exp(E_F/kT).\exp(-E_s/kT) \tag{4.33}$$

Normalizing to n_s, as in case 1, we now get

$$n_s = \left(\frac{2\pi mkT}{h^2}\right)^{3/2}\exp\frac{E_F}{kT}$$

or
$$\exp\left(-\frac{E_F}{kT}\right) = \frac{1}{n_s}.\left(\frac{2\pi mkT}{h^2}\right)^{3/2} \tag{4.34}$$

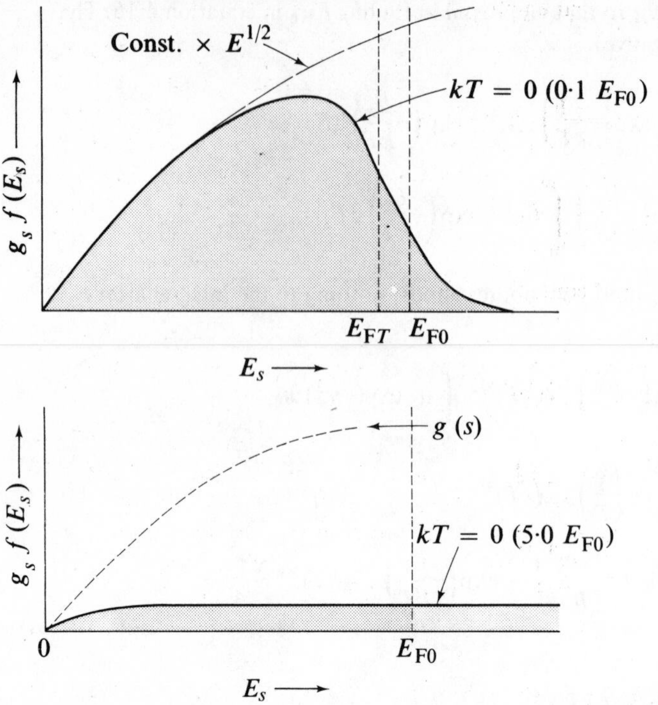

Fig. 4.4 (a) The number of occupied states and the shift in the Fermi energy, due to increasing temperature.
(b) The occupied states, drawn for an extremely high temperature, such that $kT = 5.0\,E_{F0}$.

To ensure the validity of equation 4.33, we must require $\exp(-E_F/kT) \gg 1$, since otherwise the denominator of equation 4.29 will not approximate closely to unity. Thus, the criterion for Maxwellian statistics to apply is

$$\frac{1}{n_s} \cdot \left(\frac{2\pi mkT}{h^2}\right)^{3/2} \gg 1 \qquad 4.35$$

Since the number density of electrons in a metal is many orders of magnitude greater than the number density of molecules in ideal gases at normal pressures, the Maxwell–Boltzmann distribution can be used in metals only when the temperature is very high.

4.4 Applications of Fermi–Dirac distributions

The Fermi–Dirac distribution function is at the heart of all modern solid-state physics and there are numbers of excellent textbooks dealing with these applications, of which we mention only those by Kittel[4] and by Ziman[5]. Another field in which Fermi–Dirac distributions are all-important is semiconductor physics. We shall briefly discuss some of the major problems in semiconductors a little later, but we begin the study of engineering applications by a study of electron emission.

4.4.1 Thermionic emission from pure metals

If a refractory metal, e.g. tungsten, rhenium or tantalum, is heated to a temperature a few hundred degrees below the melting point, in high vacuum, a copious electron emission can be drawn from it to a positive anode. The emission current can be in excess of 10^5 A/m^2, but this still represents the emission of only a very small proportion of the electrons incident on the surface from within the metal; thus we can calculate as though the system were in equilibrium. From the quantum-mechanical viewpoint, the heated metal can be regarded as a potential well in which the walls, instead of being infinitely high so as to allow no particle transmission, are of finite height, determined by the surface-potential barrier. Thus, some of the incident particles will have sufficient energy to surmount the surface barrier (which is called the 'surface *work function*'). These electrons are emitted. The work function, symbol ϕ, for pure metals varies between about 2.0 eV and about 6.0 eV, with tungsten having a value close to 4.5 eV. The work function of a pure-metal crystal depends on the crystal face, and ϕ values quoted for ordinary polycrystalline wires or strips are averaged over a particular arrangement of crystal faces by the experimental techniques usually employed. To recapitulate, in tungsten an electron will not be emitted unless the energy component normal to the surface is in excess of $E_F + \phi_W$ eV. Any energy in excess of this value is in the form of translational kinetic energy and so the average emitted electron has a finite velocity away from the surface, after emission.

The calculation is most easily made by transforming the energy distribution into a distribution in terms of the three component velocities u, v, w. Then, using equation 4.10, we have

$$E = \tfrac{1}{2}m(u^2 + v^2 + w^2) = \frac{\pi^2\hbar^2}{2m}\left[\frac{n^2}{a_1{}^2} + \frac{m^2}{a_2{}^2} + \frac{l^2}{a_3{}^2}\right]$$

for a parallelepiped of volume $V = a_1 . a_2 . a_3$.

Then $du = \dfrac{h}{2m} . \dfrac{dn}{a_1}$ etc. and the increment in the number of *positive* velocity states is

$$dn\,dm\,dl = \left(\frac{m}{h}\right)^3 V\,du\,dv\,dw$$

Therefore, we can write the distribution function for the number of *electrons* in $du\,dv\,dw$ as

$$dN = 2V\left(\frac{m}{h}\right)^3 \frac{du dv dw}{1 + \exp(-E_F/kT)\exp[m(u^2 + v^2 + w^2)/2kT]} \qquad 4.36$$

Suppose the surface is normal to the direction of u, then the u-directed velocity of an emitted electron is greater than

$$u_0 = \left\{\frac{2}{m}(E_F + e\phi)\right\}^{1/2} \qquad 4.37$$

but v and w can have any value between $\pm\infty$. The current element along u, correspondi to a single electron with velocity u, is $j = eu$ so the emitted current density can be written

$$J = 2\left(\frac{m}{h}\right)^3 e \int_{u_0}^{\infty} u \,.\, du \iint_{-\infty}^{+\infty} \frac{dv dw}{1 + \exp(-E_F/kT)\exp[m(u^2 + v^2 + w^2)/2kT]}$$

The double integral is easily evaluated in polar coordinates $v = \rho\cos\theta, w = \rho\sin\theta$. It then becomes

$$4\int_0^{\pi/2} d\theta \int_0^{\infty} \frac{\rho d\rho}{1 + \exp(m\rho^2/2kT)\,.\,\exp[-E_F/kT - mu^2/2kT]}$$

$$= \frac{2\pi kT}{m} \ln\left\{1 + \exp\left[\frac{E_F}{kT} - \frac{mu^2}{2kT}\right]\right\}$$

$$\therefore J = \frac{4\pi em^2kT}{h^3} \int_{u_0}^{\infty} u \ln\left\{1 + \exp\left[\frac{E_F}{kT} - \frac{mu^2}{2kT}\right]\right\} du$$

Now we have previously noted that, although E_F/kT is large, the whole exponential term must be small and we can use $\ln(1+\delta) \to \delta$ to get

$$J \approx \frac{4\pi em^2kT}{h^3} \exp\frac{E_F}{kT} \int_{u_0}^{\infty} u \exp\left(-\frac{mu^2}{2kT}\right) du$$

$$= \frac{4\pi em(kT)^2}{h^3} \exp\frac{E_F}{kT} \exp\left(-\frac{mu_0^2}{2kT}\right) \qquad 4.38$$

If we now insert the value of u_0^2 from equation 4.37, we get

$$J = \frac{4\pi em(kT)^2}{h^3} \exp\left(-\frac{e\phi}{kT}\right)$$

$$= A_0 T^2 \exp(-e\phi/kT)$$

$$= 1.204 \times 10^6 T^2 \exp(-e\phi/kT)\, A/m^2 \qquad 4.39$$

which is the so-called Richardson–Dushman equation. A_0 is called the 'Richardson constant.'

However, this equation is slightly spurious. The trouble basically arises with the definition and measurement of ϕ. Many commonly used measures of ϕ measure ϕ with respect to E_{F0}. Then, from our knowledge of the way E_F depends on temperature, we should expect ϕ to *increase* with increasing temperature, by an amount equal to $E_F - E_{F0}$. In other words, we should put

$$\exp(-mu_0^2/2kT) = \exp[-(E_{F0} + e\phi)/kT]$$

so that equation 4.39 becomes

$$J = A_0 T^2 \exp(-e\phi/kT)\exp(-\pi^2 kT/12\, E_{F0}) \qquad 4.40$$

The second exponential is essentially $\exp(-kT/E_{F0})$, which is not much less than 1. However, ϕ itself depends on temperature, because as the lattice expands the surface forces change. The net effect is that measurements fitted to equation 4.39 fail to give the expected value of the constant, whose experimental value is often only half the theoretical value. The discrepancy used to be ascribed to reflection from the surface barrier of some of the electrons which ought to have been emitted. If the reflection coefficient is R, this means that equation 4.39 ought to be multiplied by $(1-R)$. No barrier shape has ever been described which can give a large enough effect to make equation 4.39 right if the temperature coefficient is ignored. In contrast, the temperature coefficient can be measured and has values such that this explanation is now held to be basically correct.

4.4.2 Photoelectric emission

The early quantum-statistical theory of photoemission from metals, which was largely due to Fowler, regards the effect as being a surface effect, and this is often called the 'external photo-effect'. Here, the light quanta are incident on the surface of the metal. A few of the conduction electrons in the metal gain enough energy to be lifted over the work-function barrier and are emitted.

This model is somewhat defective, and it is better to regard photoemission as a volume effect. Light penetrates metal surfaces to a depth which is of the order of magnitude of one wavelength of the incident radiation, that is the intensity is reduced to about 1% in such a length. In this process the number density of photons is a decreasing function of distance from the surface, but their energy $\hbar w$ remains unchanged; thus, electrons are energized both near the surface and also from depths of a few thousand Å (1Å = 10^{-10}m). However, the mean free paths of electrons in metals are only several hundred Å, and the electrons from the lower layers thus may make several, 10 or more, collisions on their way to the surface. The conclusion is that few of the electrons actually excited are finally emitted. Also, metals reflect most of the incident light at the surface, and for these two reasons are poor photoemitters.

Semiconductors, on the other hand, absorb most of the incident light and, because the density of free electrons is much lower than in metals, the probability of reabsorption is much less. Thus, in general, semiconductor surfaces give photoemissions which are several orders of magnitude greater than those of metals and, indeed, some special semiconductors give quantum yields of 50%.

The theory we give here is over-simplified, but it serves as a first step towards a better theory. The notation is the same as in section 4.4.1. The minimum velocity for emission, u_0, after excitation by a photon of energy $\hbar\omega$ is fixed by

$$\hbar\omega + \tfrac{1}{2}mu_0^2 = E_{F0} + e\phi \qquad 4.41$$

Let α = probability that a photon actually excites an electron which is subsequently emitted. Then, following the last section, we have

$$J = \alpha \,.\, \frac{4\pi e m^2 kT}{h^3}\int_{u_0}^{\infty} u \ln\left\{1 + \exp\left[\frac{E_{F0}}{kT} - \frac{mu^2}{2kT}\right]\right\} du$$

Let $x = (1/kT)\ [\frac{1}{2}mu^2 + \hbar\omega - (E_{F0} + e\phi)]$ and $e\phi = \hbar\omega_0$. Thus, ω_0 is 2π x the Einstein threshold frequency. From equation 4.31, $x = 0$, when $u = u_0$, so the lower limit in the integral over x is zero. Or

$$J = \alpha\ \frac{4\pi e m^2 (kT)^2}{h^3} \int_0^\infty \ln\left\{1 + \exp\left[\frac{\hbar(\omega - \omega_0)}{kT} - x\right]\right\} dx$$

$$= \alpha\,.\, A_0 T^2 F(\omega - \omega_0) \qquad 4.42$$

Here, we have introduced the Fowler function defined by the integral. The evaluation of the integral is rather tedious and is relegated to Appendix 3. The result is, expressed in terms of

$$\zeta = \frac{\hbar(\omega - \omega_0)}{kT} \qquad 4.43$$

$$F(\omega - \omega_0) = \pi^2/6 + \zeta^2/2 - F(-\zeta) \qquad 4.44$$

where $F(-\zeta) = \sum \frac{(-1)^{n+1}}{n^2} \exp(n\zeta)$ 4.45

Suppose that we require to measure the photoelectric work function of a metal. The photoelectric yield is then measured as a function of ω for several temperatures. A plot of $\ln(J/A_0T^2)$ against $\hbar(\omega - \omega_0)/kT$ is made for an arbitrary value of ω_0, then the value of ω_0 is adjusted until the points fit best on the Fowler function. For a number of metals the Fowler plot gives an excellent representation of the photoelectric behaviour, over quite a large temperature range. This experimental fact must be interpreted as meaning that the coefficient α has only a weak temperature dependence. The Fowler function is plotted in Fig. 4.5. As an example of the agreement which can be achieved between theory and experiment, we quote the case of Pd. This metal has a work function of 4.97 eV and data taken over the temperature range 400–1100 K are an excellent fit on the Fowler curve.

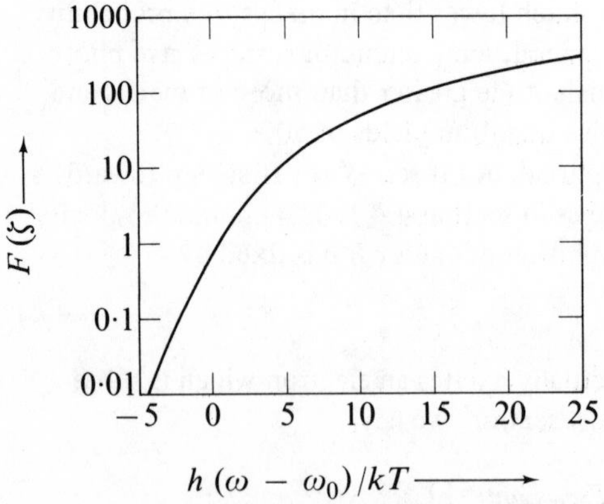

Fig. 4.5 The Fowler function giving the photo-emission from a metal for frequencies near the Einstein limit.

4.5 Semiconductors and the Fermi distribution

It is assumed that the reader has, at least, a nodding acquaintance with the band theory of solids. Individual atoms have, in quantum theory, a series of allowed discrete energy levels. When atoms are brought together, the interaction between them can be described by an extra term in the Hamiltonian. This causes a broadening of the original levels to an extent sufficient to accommodate the number of microparticles taking part in the interaction. The band structure is a characteristic of the particular material. Frequently, the broad bands overlap, but sometimes circumstances are such that electrons fill all the available levels up to the top of a band (called the valence or filled band) and there is then an energy gap before the next empty gap is reached. This situation is illustrated in Fig. 4.6. The band which at 0 K is void of electrons is called the 'conduction band'. Electrons can move freely in the conduction band. If the gap E_g is large, say > 2.0 eV, even at high temperatures very few electrons will be thermally excited into the conduction band and the material is an *insulator*. If E_g is <1.0 eV, electrons will be present, in small densities, even at ordinary temperatures and the material is called an '*intrinsic*'

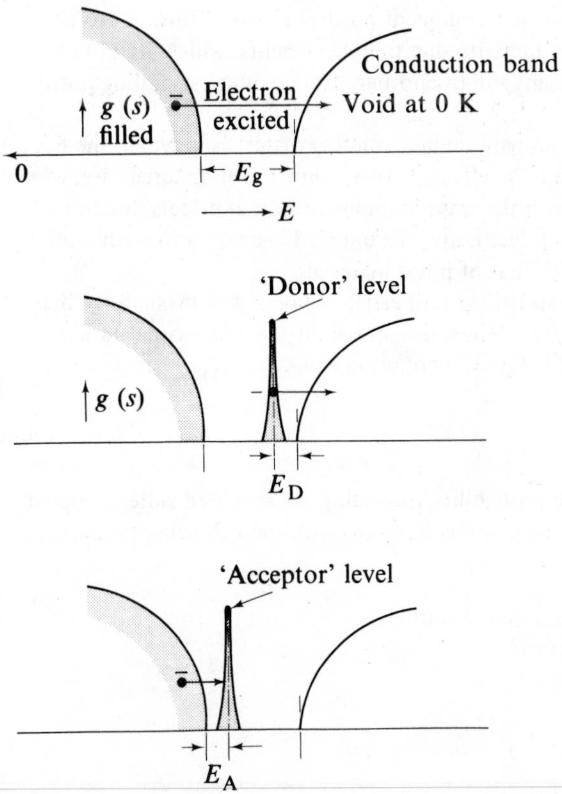

Fig. 4.6 (a) Illustrating the concept of an intrinsic semiconductor.
(b) An n-type semiconductor.
(c) A p-type semiconductor.

semiconductor. Intrinsic semiconductors show an electrical conductivity which increases with temperature, unlike poorly conducting metals. It will shortly become clear that intrinsic semiconductors must be observed in a condition of extreme purity, since even small traces of impurity modify the behaviour considerably.

Impurity semiconductors are those in which specific types of impurity atom are built into the basic material lattice in such a manner as either to produce electrons in the conduction band or alternatively to accept electrons *from* the valence band. Semiconductors of the first type are called 'n' (for 'negative') type semiconductors, while those of the second type are 'p' (for 'positive'). The reason for this nomenclature will soon be clear.

First n-type material. The donor centres are filled with electrons at 0 K. At higher temperatures, some of these electrons are excited to the conduction band and give rise to electrical conductivity of the ordinary free-electron type shown in good conductors. However, the density of free electrons increases with T and the conductivity varies in the same sense as T.

In p-type material, the acceptor centres are void at low temperatures. At higher temperatures, electrons from the valence band occupy the vacancies and leave 'holes' behind them in the valence band. Holes, or states which electrons can fill, will shortly be shown to behave as fermions of positive charge. Thus, in p-type material we can speak of a conductivity due to hole currents, which are in fact electrons moving from one vacant site to another. The holes act as mobile positive charges, whence the name.

Since it is very difficult to prepare semiconducting materials in which the doping agents are either solely donors or solely acceptors, commercial materials are, whenever possible, heavily doped with the wanted impurity so that effects due to unwanted impurities are swamped. Naturally, the detailed behaviour of such semiconductors is not identical with that of purer materials.

First, let us investigate the statistical properties of holes. The probability that a particular state is occupied is $f(E)$. Thus, the probability that the same state is unoccupied is $1 - f(E)$. Call this $f_h(E)$; then, using equation 4.7,

$$f_h(E_s) = \frac{1}{1 + \exp[(E_F - E_s)/kT]} \qquad 4.46$$

We can regard $f_h(E_s)$ as the probability of finding the specified state occupied by a hole. Then, we can describe holes as fermions with the following properties:

hole charge $= +|e|$
hole momentum $= -$ electron momentum
Fermi level $=$ constant $-E_F$
energy E_h $=$ constant $-E_s$

Using these,

$$f_h(E_s) = \frac{1}{\exp[(E_h - E_F)/kT] + 1} \qquad 4.47$$

Also, we should mention at this point that the effective masses m_e^* and m_h^* of electrons and holes respectively ought to be used in accurate calculations. The two effective masses are not the same for a given material.

4.5.1 Number densities of electrons and holes and the position of the Fermi level in an intrinsic semiconductor

The situation is shown in Fig. 4.6(a) and it is immediately clear that the number density of electrons in the conduction band, n_-, must equal the number density of holes in the valence band. We take the energy origin at the top of the valence band. Let E_c = an energy in the conduction band, E_v an energy in the valence band. Then, for a 'particle' with momentum $\boldsymbol{p}$,

$$E_v = -\frac{p^2}{2m_h^*}$$

$$E_c = E_g + \frac{p^2}{2m_e^*}$$

Here, the magnitude of the momentum vector is the important quantity, since there is no question of any directional effect. Then, we can rewrite equation 4.36 in terms of momentum and integrate over all values of momentum to obtain

$$n_- = \frac{2}{h^3}\int_0^\infty \frac{\mathrm{d}p}{\exp\beta(E_g + p^2/2m_e^* - E_F) + 1}$$

$$n_+ = \frac{2}{h^3}\int_0^\infty \frac{\mathrm{d}p}{\exp\beta(E_F + p^2/2m_h^*) + 1}$$

Here, β has been used for $1/kT$ to save writing, and equation 4.47 has been used for holes. For ordinary laboratory temperatures, as before, $E_{F_0} \gg kT$, also E_g is considerably smaller than E_{F0} so $|E_g - E_{F0}| \gg kT$. This means we can write

$$n_- \approx \frac{2}{h^3}\int_0^\infty \exp[-\beta(E_g + p^2/2m_e^* - E_F)]\ \mathrm{d}p$$

$$= 2\,(2\pi m_e^*\, kT/h^2)^{3/2}\exp[-\beta(E_g - E_F)] \qquad 4.48$$

$$n_+ \approx \frac{2}{h^3}\int_0^\infty \exp[-\beta(E_F + p^2/2m_h^*)]\ \mathrm{d}p$$

$$= 2(2\pi m_h^* kT/h^2)^{3/2}\exp(-\beta E_F) \qquad 4.49$$

Equating n_- and n_+ for electrical neutrality,

$$\exp\beta E_F = (m_h^*/m_e^*)^{3/4}\exp(\beta E_g/2)$$

$$\text{or}\quad E_F = \frac{E_g}{2} + \frac{3}{4\beta}\ln\left(\frac{m_h^*}{m_e^*}\right)$$

$$= \frac{E_g}{2} + \frac{3kT}{4}\ln\left(\frac{m_h^*}{m_e^*}\right) \qquad 4.50$$

Using equation 4.50 in equations 4.48 and 4.49, we find

$$n_- = n_+ = 2\left\{\frac{2\pi(m_e^* m_h^*)^{1/2}kT}{h^2}\right\}^{3/2} \exp\left(-\frac{E_g}{2kT}\right) \qquad 4.51$$

These results show, firstly, that the Fermi level of the intrinsic material lies almost exactly in the centre of the gap, since $\ln(m_h^*/m_e^*)$ must be very small; secondly, the number density of electrons in the conduction band (or of holes in the valence band) increases rapidly with temperature, as $T^{3/2} \exp(T^{-1})$. The exponential term is much more important than the $T^{3/2}$. Thirdly, when numbers are inserted in equation 4.51 for a particular material, it is commonly found that the electron density in the conduction band is several orders of magnitude below the density in metals. To obtain high-conductivity material it is therefore necessary to resort to doping.

4.5.2 The n-type semiconductors

Let N_D = density of donor impurity atoms
n = density of conduction electrons
f_n = distribution function for electrons, some of which are in the conduction band and some in the donor levels

The condition for electrical neutrality is

$$N_D = \int_{-\infty}^{+\infty} f_n \cdot \frac{dp}{dE}\, dE \qquad 4.52$$

For the donor levels we can express dp/dE as a delta function, i.e.

$dp/dE \approx N_D\, \delta(E + E_D)$

where we measure the energy relative to the bottom of the conduction band. Then, equation 4.52 becomes

$$N_D = \int_{-\infty}^{0} N_D\, f(E)\delta(E + E_D)\, dE + \int_{0}^{\infty} f(E)\frac{dp}{dE} \cdot dE \qquad 4.53$$

where $f(E)$ is the Fermi function.

If n is small, Maxwell–Boltzmann statistics apply and we can approximate the second integral by

$$\frac{8\pi}{h^3} \exp\left(\frac{E_F}{kT}\right) \int_0^\infty \exp\left(-\frac{E}{kT}\right) p^2\, dp$$

Equation 4.53 then becomes

$$N_D = \frac{N_D}{\exp[-(E_D + E_F)/kT] + 1} + 2\left(\frac{2\pi mkT}{h^2}\right)^{3/2} \exp\frac{E_F}{kT} \qquad 4.54$$

whence $$N_D\left(\frac{\exp[-(E_D + E_F)/kT]}{\exp[-(E_D + E_F)/kT] + 1}\right) = 2\left(\frac{2\pi mkT}{h^2}\right)^{3/2} \exp\frac{E_F}{kT}$$

If $(E_D + E_F/kT) \gg 1$, at low temperatures, this can be approximated by

$$N_D \exp[-(E_D + E_F/kT)] = 2(2\pi mkT/h^2)^{3/2} \exp(E_F/kT)$$

Solving for E_F we get

$$\text{a)}\quad E_F \approx -\frac{E_D}{2} + \frac{kT}{2}\ln\frac{N_D}{2(2\pi mkT/h^2)^{3/2}} \qquad 4.55$$

Alternatively, if $\dfrac{N_D}{(2\pi mkT/h^2)^{3/2}}\exp\left(\dfrac{E_D}{kT}\right) \ll 1$,

$$\text{b)}\quad E_F \approx kT\ln\frac{N_D}{2(2\pi mkT/h^2)^{3/2}} \qquad 4.56$$

Equations 4.55 and 4.56 are two limiting cases.
The distribution functions f_n are

$$\text{Case 1}\quad f_{n1} = \frac{N_D}{2(2\pi mkT/h^2)^{3/2}}\exp\left[-\frac{(E_D + 2E)}{2kT}\right] \qquad 4.57$$

$$\text{Case 2}\quad f_{n2} = \frac{N_D}{2(2\pi mkT/h^2)^{3/2}}\exp\left(-\frac{E}{kT}\right) \qquad 4.58$$

which can be written more compactly by putting $(2\pi mkT/h^2)^{3/2} = N_c$.
Then, from equation 4.57,

$$f_{n1} = \frac{N_D}{2N_c}\exp\left(-\frac{E_D}{2kT}\right)\exp\left(-\frac{E}{kT}\right)$$

$$= f_{n2}\exp\left(-\frac{E_D}{2kT}\right) \qquad 4.59$$

Thus, as the temperature increases, the number of electrons in the conduction band increases exponentially from zero at $T = 0$ K, towards a constant value at high temperatures for which $\exp(-E_D/2kT) \to 1$. Physically, this means that all the donors have ionized and have contributed their electrons to the conduction band.

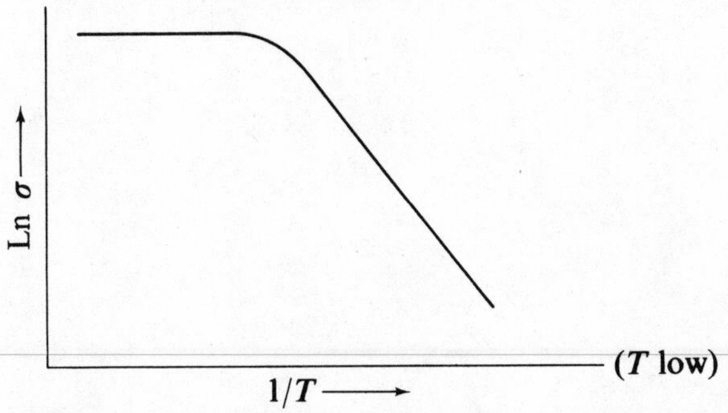

Fig. 4.7 Conductivity vs. temperature for an n-type semiconductor.

Later in this book we show that the electrical conductivity of a semiconductor is given by

$$\sigma = (e^2\tau/m_e^*)\, n$$

Thus, the conductivity varies as shown in Fig. 4.7. A detailed consideration shows that E_D can be deduced from the linear variation of ln σ at low temperatures. N_D is set by the extent of the doping employed and is thus a material constant which is (more or less) under control. If N_D and E_D are known, the above theory allows the detailed behaviour to be calculated.

For acceptor impurities, the same results apply, replacing N_D by N_A and $E_D = E_A$. In general, it is possible to make materials in which E_D (or E_A) is a fraction of the gap energy E_g. Thus, extrinsic semiconductors can be made with relatively high values of conductivity. In practical materials, it is commonly the case that both donor and acceptor impurities are present in the same material or, even more complicated, there may be donors and acceptors of more than one species. The theory can be extended to these cases without any difficulties of principle, but the expressions become extraordinarily cumbersome and we shall not make the extension here.

5

Quantum Statistics: the Bose–Einstein Distribution

Certain microparticles, He^4 atoms for example, have integral values of the total spin angular momentum, viz. 0, $\hbar$, $2\hbar$, . . . , etc. This means that the total wave function is symmetric, that is the wave function remains unchanged when two particles are interchanged, the spins being interchanged as well as the space coordinates. Such particles are called 'bosons' and their behaviour differs markedly from that of fermions, whose wave function changes sign on interchange of particles.

In the case of bosons it is clear that, if the wave function is unchanged by interchange of particles, there can be no way of distinguishing the particles from one another, confirming the original hypothesis. Moreover, the Pauli exclusion principle no longer operates and there is no restriction on the number of particles in a particular state. An important 'particle' which obeys Bose–Einstein statistics is the photon. Photons are peculiar among bosons in that their number is not conserved. If we consider a cavity which is filled with electromagnetic radiation spread over all the resonant frequencies, we can also describe the system as an isolated box full of bosons. The photons collide with the walls but, unlike gas atoms or electrons, they are not all reflected, some being absorbed instead. The requirement that the number of particles present remains constant must be relaxed, and the only condition which operates is that the total energy is conserved in an isolated system.

5.1 Simple derivation of the Bose–Einstein distribution

The notation is the same as that of section 4.1. The number of states is $g(s)$. A particular state has energy E_s and there are n_s particles in that state. We must calculate the number of ways in which the n_s particles can be distributed among the $g(s)$ states. The problem is identical with the following problem: n_s balls are to be arranged in $g(s)$ boxes, allowing any number of balls in a box. The mental procedure is to lay out the $n_s + g(s)$ objects – balls and boxes – in a line in random order, achieved say by tossing a penny. The arrangement is chosen to start with a box and might continue

Box (Ball Ball Ball) Box (Ball Ball) Box . . .

Then, move those balls which are immediately to the right of a given box into that box. For the arrangement given, there are three balls in box 1, two in box 2 and so on, in random fashion. Since $(n_s + g(s) - 1)$ objects move, there are $(n_s + g(s) - 1)!$ such arrangements, but many of these arrangements are identical; for example, we must divide by 3! to take account of the fact that the balls in box 1 are all identical. There are $n_s!$ such permutations. Similarly, there are $(g(s) - 1)!$ permutations of the boxes. Finally we get

$$W = \frac{(n_s + g(s) - 1)!}{n_s!\,(g(s) - 1)!} \qquad 5.1$$

as the number of distinguishable arrangements. Then, summing over all values of s, we get

$$W = \prod_s \frac{(n_s + g(s) - 1)!}{n_s!\,(g(s) - 1)!} \qquad 5.2$$

Since $g(s)$ is a large number in all normal cases, compared with both n_s and unity,

$$W = \prod_s \frac{(n_s + g(s))!}{n_s!\,(g(s) - 1)!} \qquad 5.3$$

For the moment we leave photons out of consideration and maximize equation 5.3 subject to conservation of numbers of bosons and of energy. Then, following the method used before, we find

$$n_s = \frac{g(s)}{\exp[(E_s - \mu)/kT] - 1} \qquad 5.4$$

after using the argument that Bose–Einstein statistics must be identical with Maxwellian for large enough energies. The quantity μ/kT replaces the multiplier α. From equation 5.4 we see that the probability distribution function for bosons is

$$f_B(E_s) = \frac{1}{\exp[(E_s - \mu)/kT] - 1} \qquad 5.5$$

Since $f_B(E_s)$ cannot exceed unity, we at once conclude that $E_s - \mu \geqslant 0$ for *all* E_s. Thus, μ is either zero or negative. Looking back over the derivation, we see that, *for photons,* since N is *not* conserved, we need only a single Lagrangian multiplier and *μ is zero.* For other bosons, μ is negative. The behaviour of equation 5.5 as a function

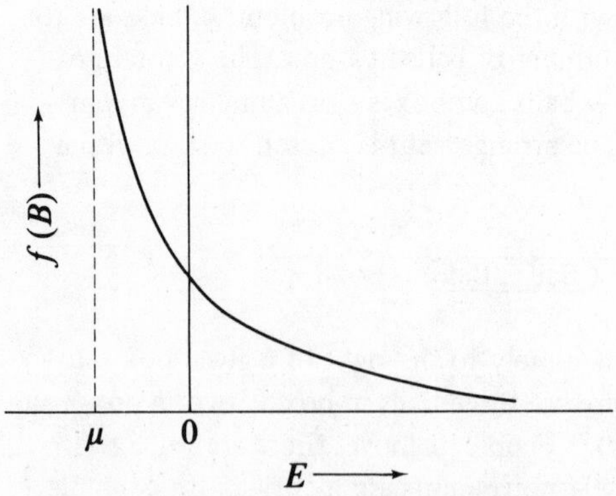

Fig. 5.1 The behaviour of the Bose-Einstein function.

of E_s is indicated in Fig. 5.1. $f_B(E_s)$ varies from a finite value to zero as E_s increases, since the single particle energy E_s is > 0. The reader will readily show that, at high energies, $f_B(E_s)$ is the same shape as the Maxwellian.

Before continuing with another derivation of the distribution, it is worthwhile drawing attention to a peculiarity which can easily be demonstrated for the case $\mu = 0$, but assuming that a fixed number of particles are present. Using the standard expression for $g(s)$, we have

$$N = \frac{(2m^3)^{1/2}}{\hbar^3\pi^2}\int_0^\infty \frac{E^{1/2}\,\mathrm{d}E}{\exp(E/kT)-1} \qquad 5.6$$

$$= \frac{(2m^3)^{1/2}}{\hbar^3\pi^2}\int_0^\infty \frac{E^{1/2}\,\mathrm{d}E}{\exp(E/kT)\,\{1-\exp(-E/kT)\}}$$

$$= \frac{(2m^3)^{1/2}}{\hbar^3\pi^2}\int_0^\infty \mathrm{d}E\,.\,E^{1/2}\exp\left(-\frac{E}{kT}\right)\left\{1+\exp\left(-\frac{E}{kT}\right)+\exp\left(-\frac{2E}{kT}\right)+\text{etc.}\right\}$$

$$= \frac{(2m^3)^{1/2}}{\hbar^3\pi^2}\int_0^\infty \mathrm{d}E\,.\,E^{1/2}\left\{\exp\left(-\frac{E}{kT}\right)+\exp\left(-\frac{2E}{kT}\right)+\exp\left(-\frac{3E}{kT}\right)+\text{etc.}\right\}$$

Put $nE/kT = x^2$, $(n/kT)\,\mathrm{d}E = 2x\,\mathrm{d}x$ and the integrals all become

$$2\left(\frac{kT}{nm}\right)^{3/2}\int_0^\infty \mathrm{d}x\,.\,x^2\exp(-x^2) = 2\left(\frac{kT}{nm}\right)^{3/2}\,.\,\frac{\sqrt{\pi}}{4}$$

The resulting integral sum is

$$(kT)^{3/2}\,.\,\frac{\sqrt{\pi}}{2}\sum_{n=1}^{\infty}\frac{1}{n^{3/2}}$$

The series converges rapidly and it is easy to see that the sum is close to 2.6. N is then given by

$$N = 2.31\,(mkT/2\pi^2\hbar^2)^{3/2} \qquad 5.7$$

Proceeding in the same way, the average energy is proportional to

$\displaystyle\sum_{n=1}^{\infty}\frac{1}{n^{5/2}}$, which is even more convergent, to 1.34. Then

$$\langle E\rangle = 2.01\,(m/2\pi\hbar^2)^{3/2}\,(kT)^{5/2} \qquad 5.8$$

It is obvious that negative values of μ will ensure that equations 5.7 and 5.8 represent *maximum* values of N and $\langle E\rangle$; thus, equation 5.7 gives the maximum number of particles which can be accommodated by a Bose–Einstein system at T K. The interesting question arises, what happens when we cool N particles down through the temperature T? The answer is that some of the particles must go into a state of zero energy which makes no contribution to the integrals. This phenomenon is called Bose–Einstein 'condensation', by analogy with the condensation of a saturated vapour. It can occur only at temperatures of a few kelvin and is important in the theories of superfluidity and superconductivity. The point we have established is that the Bose–Einstein distribution behaves very differently from the Fermi–Dirac, as regards the comparison between E_{F0} and μ.

5.2 Derivation from the grand canonical ensemble*

The average number of particles in a box or sub-system is again given by equation 4.19, viz.

$$\langle n_r \rangle = kT \frac{\partial}{\partial \mu} \ln \sum_{n_r} \exp\left(\frac{\mu - E_r}{kT}\right)^{n_r} \Omega(n_r) \qquad 5.9$$

As in section 4.2, $\Omega(n_r)$ = unity. Therefore

$$\langle n_r \rangle = kT \frac{\partial}{\partial \mu} \ln \sum_{n_r=0}^{N} \exp\left(\frac{\mu - E_r}{kT}\right)^{n_r}$$

If $\exp[(\mu - E_r)/kT)] < 1$, the series converges and, moreover, it will converge sufficiently fast so that we can let $N \rightarrow \infty$. Then,

$$\langle n_r \rangle = kT \frac{\partial}{\partial \mu} \ln \sum_{n_r=0}^{\infty} \exp(-\alpha)^{n_r}$$

$$= kT \frac{\partial}{\partial \mu} \ln \{1 - \exp(-\alpha)\}^{-1}$$

$$= \frac{1}{\exp[(E_r - \mu)/kT] - 1} \qquad 5.10$$

Equation 5.10 is in precise agreement with equation 5.5 for $E_r = E_s$. Once again we have the condition $\exp[(\mu - E_s)/kT] < 1$ for all E_s, and μ must therefore be negative.

5.3 Remarks on the Bose–Einstein distribution at very high energies and normal temperatures

We can write the Bose–Einstein and Fermi–Dirac distributions in general form as $f = \exp[E - \mu)/kT] \pm 1$, where the top sign is Fermi–Dirac and the bottom is Bose–Einstein. Clearly, when E is large with respect to kT and μ is zero or negative, the exponential is large. Then, the Bose–Einstein distribution becomes Maxwellian –

*This section may be omitted on a first reading.

just as was the case for the Fermi–Dirac; in other words, the sign of the unity in the denominator makes no difference when the exponential is large.

When the distribution is applied to particles whose number is conserved, we have the usual relationships

$$N = \sum_r \frac{1}{\exp[(E_r - \mu)/kT] - 1} \tag{5.11}$$

and $$E = \sum \frac{E_r}{\exp[(E_r - \mu)/kT] - 1} \tag{5.12}$$

These can be expanded as in section 5.1 when $\mu < 0$. The reader should verify for himself that, for example, the energy density E/V is given by

$$\frac{E}{V} = \frac{3}{2}(kT)\left(\frac{2\pi mkT}{h^2}\right)^{3/2} \sum_{n=1}^{\infty} \frac{\exp(n\mu/kT)}{n^{5/2}} \tag{5.13}$$

5.4 Applications of the Bose–Einstein distribution function

We shall study the applications of the Bose–Einstein distribution function in relation to electromagnetic radiation, i.e. applications to photon statistics, which has engineering applications in the theory of optical temperature measurement, the theory of masers and lasers, and the theory of noise in optical-communications. Another type of boson, which we shall briefly discuss, is the paired electron state, called a 'Cooper pair' which is basic in superconductivity theory.

5.4.1 Bose–Einstein statistics and the Planck radiation formulae

Planck's radiation formulae, in which quantum ideas were introduced for the first time, were derived on the basis of *ad hoc* assumptions of a very bold nature. Although Planck's results gained rapid acceptance because they were in remarkably good agreement with experiment, it was still extremely gratifying that, about ten years later, the same result was derived from the Bose–Einstein distribution.

The physical system to be considered consists of a cavity resonator, isolated from the outside world except for a very small hole through which some radiation is emitted for observation purposes. The resonator is at temperature T K. As stated, in the introductory part of this chapter, photons incident on the walls are not invariably re-emitted into the resonator; thus, the number of photons is *not* conserved and we must abandon the condition $\Sigma n_s = N$. We also no longer require the Lagrangian multiplier μ, and can equate it to zero. Thus, for photons,

$$n_s = \frac{g(s)}{\exp \beta E_s - 1} \tag{5.14}$$

We very nearly solved the problem of evaluating $g(s)$ for photons when we discussed the Schrödinger probability waves in a rectangular piece of metal in section 4.0, where we found the number of probability waves in a box of volume V was $VK^2\,\mathrm{d}K/2\pi^2$ in a range $\mathrm{d}K$ at K. The eigenfunctions for a rectangular electromagnetic resonator are of exactly the same form as those for the solution of the Schrödinger equation for a metal block of the same dimensions, so the same result holds. We pointed out in that discussion that the K's were analogous to the propagation constants of e.m. waves, which are defined by

$$K = 2\pi/\lambda = 2\pi\nu/c = \omega/c \qquad 5.15$$

where c = velocity of light.

We must, however, remember that each eigenfunction is associated with both a magnetic field and an electric field, so that in classical theory we should associate the energy $kT/2$ with the E mode and another $kT/2$ with the H mode. Thus, we must allow for this phenomenon of polarization by multiplying by a factor of 2 in $g(s)$. Then, $g(s)$ becomes

$$g(s) = \frac{V\,.\,8\pi\nu^2\,\mathrm{d}\nu}{c^3} \qquad 5.16$$

$$\text{and } n_{\mathrm{s}} = \frac{V\,.\,(8\pi\nu^2/c^3)\,\mathrm{d}\nu}{\exp(\hbar\omega/kT) - 1} \qquad 5.17$$

since the energy of the photon frequency V is $h\nu$ or $\hbar\omega$. The energy of the n_{s} photons is obtained by multiplying equation 5.17 by $\hbar\omega$. The energy density in $\mathrm{d}w$ at w is then $n_{\mathrm{s}}\hbar\omega\,V^{-1}$. Call this energy density $\rho(\omega_1\,T)\,\mathrm{d}\omega$; we then find

$$\rho(\omega, T)\,\mathrm{d}\omega = \frac{\hbar\omega^3\,\mathrm{d}\omega}{\pi^2 c^3\{\exp(\hbar\omega/kT) - 1\}} \qquad 5.18$$

Equation 5.18, which is precisely the same as Planck's formula, is plotted in normalized form in Fig. 5.2.

The total energy in the cavity is obtained by integrating equation 5.18 from $\omega = 0$ to $\omega \to \infty$; i.e.

$$E_{\mathrm{T}} = \frac{\hbar}{\pi^2 c^3}\int_0^\infty \frac{\omega^3\,\mathrm{d}\omega}{\{\exp(\hbar\omega/kT) - 1\}}$$

If we put $\hbar\omega/kT = x$, this becomes

$$E_{\mathrm{T}} = \frac{\hbar}{\pi^2 c^3}\left(\frac{kT}{\hbar}\right)^4 \int_0^\infty \frac{x^3\,\mathrm{d}x}{\exp x - 1}$$

The integral is standard and has the value $\pi^4/15$; therefore

$$E_{\mathrm{T}} = \frac{\pi^2 (kT)^4}{15(\hbar c)^3} = \sigma T^4 \qquad 5.19$$

The constant σ is called 'Stefan's constant'. Since equation 5.19 describes a measurable quantity, if k and c are accurately known, this result can be used to determine $\hbar$, and was so used before better techniques were evolved. The equation is also basic in the theory of cooling of hot surfaces and in optical pyrometry.

We can now use equation 5.19 and the thermodynamic relationships of Appendix 2 to find the radiation pressure, that is the pressure exerted on the walls of the cavity by the incident photons.

The energy in the volume V is $E = \sigma VT^4$. Using equation A2.5 of Appendix 2,

$$F = -T\int \frac{E\mathrm{d}T}{T^2} = -\frac{1}{3}\sigma VT^4 \qquad 5.20$$

But, from equation A2.4, $p = -\dfrac{\partial F}{\partial V} = \dfrac{1}{3}\sigma T^4$

$$= \frac{1}{3}\frac{E}{V} \qquad 5.21$$

This expression can also be obtained by evaluating the interchange of energy between the photons and the walls. Also, it can be calculated by electromagnetic theory. The present derivation is extremely compact by comparison.

Before leaving the Planck formula, we ought to mention that the results of this section are derived with respect to a false zero. This makes no difference in the present application, but when we discuss optical noise, later on, the distinction acquires real importance.

The energy levels of quantized simple harmonic oscillations are given by

$$E_n = (n + \tfrac{1}{2})\,\hbar\omega \qquad 5.22$$

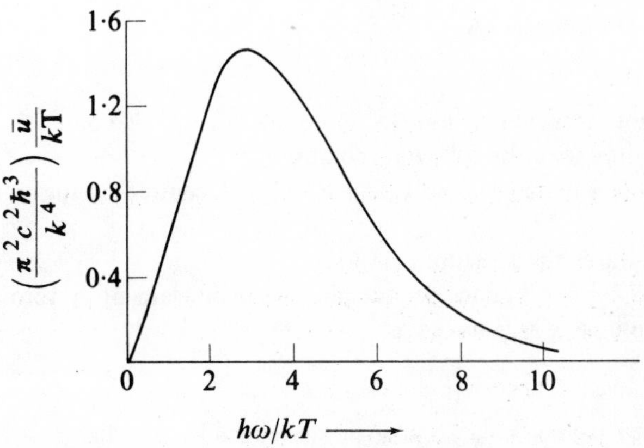

Fig. 5.2 Notation for the calculation of radiation through a small aperture.

Carrying out the energy summation for N_B bosons we find

$$E_T = N_B \left\{ \frac{\hbar\omega}{2} + \frac{\hbar\omega}{\exp(\hbar\omega/kT) - 1} \right\} \tag{5.23}$$

Thus, the earlier calculations ignore the zero-point energy $N_B\hbar\omega/2$, which is important only at low temperatures and high (optical) frequencies.

Conversely, for laboratory temperatures and radio and microwave frequencies,

$$\exp(\hbar\omega/kT) \to 1 + \hbar\omega/kT + \ldots$$

and $\quad E_T \to N_B kT \qquad$ 5.24

which is the classical result for N_B oscillators with two degrees of freedom and energy $kT/2$ per degree of freedom.

5.4.2 The radiation through a small aperture in the cavity wall

For observation purposes, we wish to calculate the radiation emitted through a small hole in cavity walls. The cavity near the aperture and the notation are shown

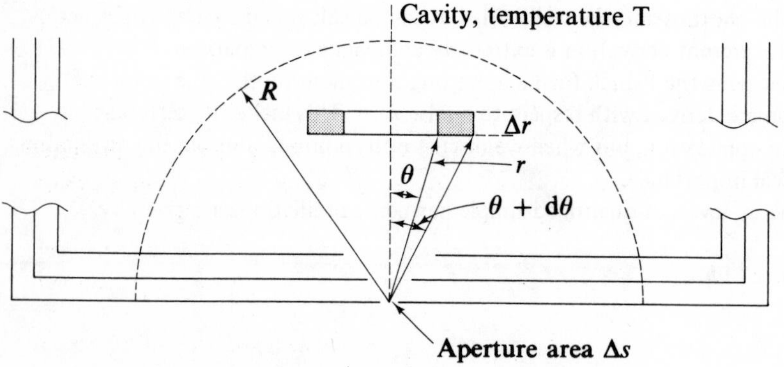

Fig. 5.3 Normal form of Planck's radiation law.

in Fig. 5.3. Consider the photons which are emitted in a short time Δt. To be emitted, a photon must comply with the following conditions:

1. it must be located inside a hemisphere of radius $R = c \,.\, \Delta t$, centred on the aperture centre;

2. it must be directed towards the aperture area Δs.

To determine the probability of (2), consider the photons in an elementary toroid of section $r\mathrm{d}\theta\,\mathrm{d}r$ at r. The volume of this toroid is

$$2\pi r \sin\theta \,.\, r\mathrm{d}\theta\,\mathrm{d}r$$

and elements of this toroid all see the same solid angle subtended by Δs, that is

$$\Delta\Omega = \Delta s \cos\theta / r^2$$

However, all the photon directions are in the solid angle $\Omega = 4\pi$ steradians, and the probability of a photon having the right direction is

$$\Delta\Omega/\Omega = \Delta s \cos\theta/4\pi r^2$$

Then, the energy in $d\omega$ at ω, which leaks from the toroid in time Δt, is

$$(E(\omega)\Delta s/2)\cos\theta\sin\theta\,d\theta\,dr\,d\omega$$

To obtain the total we must integrate from $r = 0$ to $r = R = c\Delta t$ and from $\theta = 0$ to $\theta = \pi/2$. The result is

$$(c\,E(\omega)/4)\,\Delta t\,\Delta s\,d\omega$$

or the energy per unit time per unit area is

$$(c\,E(\omega)/4)\,d\omega$$

But we already know how to integrate over ω to get the total energy, and we can therefore write the final answer as

$$\pi^2(kT)^4/60c^2\hbar^3 \text{ per unit area}$$

Clearly, for the hole to be 'small', this leakage must be much less than the total stored energy. Measurement can be made on the spectral energy distribution over a wide range of frequencies and it was these measurements, which disagreed with the Wien and Rayleigh–Jeans laws, which confirmed Planck's formula.

5.4.3 Optical pyrometry

A technically important use of these results is in optical pyrometry, the measurement of the temperature of incandescent bodies by optical methods. In the most common type of optical pyrometer, which is called the 'disappearing-filament' pyrometer, an optical system is used to produce an image of the hot body in the plane of a heated tungsten filament. An eyepiece is used to view the filament and the optical image. The adjustment of the filament temperature, by control of the heating current until the filament is no longer visible against the image of the hot body, can be done with surprising accuracy. Pyrometers reading from just below 700 °C to 2500 °C can be realized in this fashion and are basically calibrated, over a narrow range of red frequencies, against a black-body radiator. However, the instrument is not normally used on black bodies but on ordinary heated objects. What is the relationship between the optical (or brightness) temperature of the object and the black-body temperature used in calibration?

The spectral emissivity of a source is defined as the following ratio:

$$\alpha = \frac{\text{emissive power of source in } d\lambda}{\text{emissive power of black body in } d\lambda}$$

where λ is the wavelength. In the situations where these pyrometers are used, $\exp(\hbar\omega/kT) \gg 1$ and we can use Wien's approximation for $E(\lambda)$, viz.

$$E(\lambda) = \frac{8\pi hc}{\lambda^5}\exp\left(\frac{-hc}{\lambda kT}\right)$$

If the pyrometer reads T_B (for brightness) and the true temperature of the body is T, we have

$$\alpha \cdot \frac{8\pi hc}{\lambda^5} \exp\left(-\frac{hc}{\lambda kT}\right) = \frac{8\pi hc}{\lambda^5} \exp\left(-\frac{hc}{\lambda kT_B}\right)$$

or

$$\ln \alpha = \frac{hc}{\lambda k}\left(\frac{1}{T} - \frac{1}{T_B}\right)$$

$$\therefore T = \frac{T_B}{1 + T_B \ln \alpha \cdot \lambda k/hc} \qquad 5.25$$

Since $\alpha < 1$, $\ln \alpha$ is negative and $T > T_B$ by an amount which is not negligible for most materials. Thus, determinations of α are prerequisite for the accurate use of optical pyrometry.

5.4.4 The Rayleigh–Jeans law

When we come to discuss the relationship between optical and electrical noise, it will be helpful to speak of the Rayleigh–Jeans law. This was derived using classical equipartition theory some years before Planck's work; nowadays, we can simply treat it as the limiting case of Planck's formulae when quantum effects are negligible, that is when $\hbar\omega \ll kT$. Then, $\exp(\hbar\omega/kT) \approx 1 + \hbar\omega/kT$ and equation 5.18 reduces to

$$\rho(\omega, T)_{[\text{Rayleigh–Jeans}]} = \frac{kT}{\pi^2 c^3} \cdot \omega^2 \, d\omega \qquad 5.26$$

Equation 5.26 therefore coincides with the initial steep part of Fig. 5.3.

The important point to be made is the following. If we consider that the ambient temperature is 300 K, $\hbar/kT \approx 2.5 \times 10^{-14}$. To make $\hbar\omega/kT = 1$, the frequency must be about 7000 GHz, or, to make the Rayleigh–Jeans approximation valid, the frequency must be $< O(700 \text{ GHz})$. This frequency is above the highest microwave frequencies used today, so we conclude that quantum effects are negligible in radio engineering *at ambient temperatures.* However, if we consider *cooled* circuits, at liquid-helium temperature, 4.2 K, then the frequency condition becomes $f < 7$ GHz, which is an ordinary microwave communication and radar frequency.

For optical frequencies, $\hbar\omega/kT > 1$, and quantum effects are predominant. On very general grounds we must therefore expect very different noise behaviour in optical systems from that in radio engineering.

5.4.5 The Einstein transition probabilities and the simple theory of the maser

This section deals with a system which is not really in equilibrium and which therefore logically ought to be discussed by kinetic theory. However, the departure from equilibrium is small and the original treatment, due to Einstein, is in the spirit of the kinetic calculations of Chapter 2.

Einstein considered that a large number of microscopic systems (gas atoms) were in overall thermal equilibrium in a microwave cavity at temperature T.

In addition, the cavity was excited by electromagnetic waves whose frequencies covered the range in which excitations of the atoms could take place. Thus, the overall problem was one of interchange of energy between photons and particles.

The number of particles in a given energy state is given by the canonical distribution, while the photons obey Bose–Einstein statistics. First, consider the particles alone, in statistical equilibrium at T K. The density of particles in states 1 and 2 are N_1 and N_2, and the probability that a transition from state 1 to state 2 takes place in Δt is $w_{12}\,\Delta t$. Then

$$-\,\mathrm{d}N_1/\mathrm{d}t = w_{12}N_1 - w_{21}N_2 = 0$$

$$-\,\mathrm{d}N_2/\mathrm{d}t = w_{21}N_2 - w_{12}N_1 = 0$$

These equations are solved by $w_{12} = w_{21}$, $N_1 = N_2$, as one would expect. When the radiation is present, the whole resonator is in equilibrium but the transition probabilities are modified because some transitions are aided by the radiation while others are hindered. The transition probability w_{12} now means the probability of moving from [particle in state 1, cavity in state E] to [particle in state 2, cavity in state $E-(E_2-E_1)$], where E_1, E_2 are the energies of the particle states. Since it is less probable that a particle will be in the higher energy state, we can put

$$w_{12} = w_{22}\exp[-(E_2-E_1)/kT] \qquad 5.27$$

from the Boltzmann theorem or, in strict equilibrium,

$$N_{2\,(\mathrm{eq.})} = N_1\exp\,[-(E_2-E_1)/kT] \qquad 5.28$$

Now, the radiation causes transitions which can be enumerated by

$$w_{12} = B_{12}\,\rho(\omega,\,T)$$

where B_{12} = proportionality constant to be found

$\rho(\omega,\,T)$ = energy density from equation 5.18 renormalized so as to give the total radiation energy

Then $\mathrm{d}N_1/\mathrm{d}t = -\,B_{12}\,\rho(\omega,\,T)\,N_1$

From the higher energy state, atoms can make spontaneous returns to the lower state and the number doing so is proportional to w_{21}, say by A_{21}. In addition, the radiation will be partly re-emitted in inverse transitions. This is called ‘induced’ radiation. Thus, we have

$$\mathrm{d}N_2/\mathrm{d}t = -\,[A_{21}N_2 + B_{21}\,\rho(\omega,\,T)\,N_2]$$

Therefore, in quasi-equilibrium with the radiation present,

$$\mathrm{d}N_2/\mathrm{d}t = \mathrm{d}N_1/\mathrm{d}t$$

or $$\frac{N_2}{N_1} = \frac{B_{12}\,\rho(\omega,\,T)}{A_{21} + B_{21}\,\rho(\omega,\,T)} \qquad 5.29$$

But, using equation 5.28, we also have

$$\frac{N_2}{N_1} = \exp[-(E_2 - E_1)/kT] \qquad 5.30$$

therefore

$$A_{21} = \rho(\omega, T)\ \{B_{12} \exp[(E_2 - E_1)/kT] - B_{21}\} \qquad 5.31$$

Clearly, only transitions in which $E_2 - E_1 = \hbar\omega$ are allowed by quantum selection rules. Then, if we put $B_{12} = B_{21}$, and insert the expression for $\rho(\omega, T)$, we finally arrive at

$$\frac{A_{21}}{B_{12}} = \frac{\hbar^3 \omega^3}{\pi^2 c^3} \mathrm{d}\omega \qquad 5.32$$

It can be argued, from detailed balancing, that B_{12} must, in fact, equal B_{21}; then equation 5.32 is Einstein's result for the transition probabilities. The spontaneous transitions are *improbable* by comparison with the induced ones if $\hbar\omega^3 \ll \pi^2 c^3$. Since these numbers become equal for frequencies $0(10^{20}$ Hz), the inequality is preserved for all radio and optical frequencies.

Let us now calculate the equilibrium differences between the population densities N_1 and N_2. From equation 5.30 we can put $N_2 = N_1 \exp(-\hbar\omega/2kT)$. But $N_1 + N_2 = N$, a constant. A little algebra then gives the following result:

$$N_1 - N_2 = N \tanh(\hbar\omega/2kT) \qquad 5.33$$

$$\approx N.\, \hbar\omega/2kT \qquad 5.34$$

which follows if $\hbar\omega/2kT \ll 1$. These equations give the population difference after the radiation has been switched on for long enough for equilibrium to be reached.

Now consider the time interval between the instant of switching on the radiation and the attainment of equilibrium. If we ignore spontaneous transitions, according to equation 5.32, we can write modified rate equations of the following form:

$$\mathrm{d}n_1/\mathrm{d}t = -w_{12} n_1 + w_{21} n_2 \qquad 5.35$$

$$\mathrm{d}n_2/\mathrm{d}t = -w_{21} n_2 + w_{12} n_1 \qquad 5.36$$

or, if $n = n_1 - n_2$ and $w_{12} = w_{21}$,

$$\mathrm{d}n/\mathrm{d}t = -2\, w_{12} n \qquad 5.37$$

or $\quad n = n_0 \exp(-2\, w_{12} t)$

But $n_0 = (N_1 - N_2) = N_1\ \{1 - \exp(-h\omega/kT)\}$

therefore $n = N_1\ \{1 - \exp(-\hbar\omega/kT)\} \exp(-2\, w_{12} t)$ 5.38

We can clearly interpret $2w_{12}$ as an inverse relaxation time τ, so this is equivalent to

$$n = N_1\ \{1 - \exp(-\hbar\omega/kT)\} \exp(-t/\tau) \qquad 5.39$$

but w_{12} depends on the power level, as well as on the characteristics of the system.

The overall behaviour is as follows. When the radiation is switched on, the system is driven into a state where the population of higher energy state is greater than the equilibrium value. A new 'equilibrium' is re-established when the number of newly created higher energy states just equals the (modified) number of downward transitions. The total energy absorbed is $(N_1 - N_2)\hbar\omega\, w_{12}$.

The operation of, for example, gas masers depends on the fact that the downward transition can take place in stages corresponding to energies $\hbar\omega_1$ and $\hbar\omega_2$. Thus, in principle, power can be supplied to the system at one frequency and then be given out at one or more different frequencies.

5.4.6 A note on negative temperature

A good deal of the reasoning of Chapter 3 depends on the fact that equilibrium statistics demands that $T > 0$; otherwise we could not justify many statements made about relative probabilities of occupation, etc.

In the last section, we have seen that by a particular arrangement we can realize a situation in which the population of a high energy state is much increased above its equilibrium value and, for suitably high levels of radiation, can exceed the population density of the *lower* energy state. Some authors find it convenient to describe this state of affairs by a Boltzmann equation with a *negative* temperature. For example, one can make statements such as the following: 'When the temperature of the transition 1–2 changes from positive to negative, the system ceases to absorb power but instead emits power.' Of course, quite analogous statements follow from the actual population difference, which changes sign too; thus, it is largely a matter of taste whether negative temperatures are introduced. However, the usage is becoming quite common and the reader ought to know what it means.

5.5 Bose statistics and superconductivity

Recent advances in materials technology have led to the production of superconducting alloys which are capable of retaining the superconducting state at spectacularly high values of magnetic field, in excess of 200 T. These discoveries will perhaps introduce revolutionary changes in high-power electrical engineering. A discussion of the theory of superconductivity demands an advanced knowledge of quantum mechanics and is quite outside the scope of this book; however, the importance of the subject warrants a brief qualitative review of some important ideas which are linked to the phenomenon of Bose–Einstein condensation which we have already discussed.

The fundamental idea which explains the superconducting state is that, under special circumstances which exist in real crystalline lattices, two electrons can be weakly bound in a pair state. They are then known as 'Cooper' pairs. In a very rough qualitative way, a lattice of atomic nuclei can be regarded in a manner reminiscent of the Debye sphere of plasma physics, and the electrons interact with one another in such a way that the coulomb repulsion cancels and leaves a small binding energy. Two electrons bound thus exhibit a symmetrical wave function – which is exactly the property which identifies the particle as a boson. We may, then, regard Cooper pairs as bosons. It turns out that the number of pairs is not conserved, so the chemical potential μ approaches zero. According to our account of Bose–Einstein condensation, the boson gas in the metal goes through a transition at some low temperature, into a state where the normal electrons are in fixed energy states and only the bosons can play a role in the conduction process. The conduction process then depends on a *three-fluid* model, the three fluids being (a) the unpaired electrons, obeying Fermi–Dirac statistics, (b) the uncondensed Cooper pairs and (c) the condensed Cooper pairs. Then, the total number of electrons is $N = N_a + N_b + N_c$. When $T > T_c$, the transition temperature, $N_c \to 0$ and $N_b \ll N_a$, so that the behaviour is that of an ordinary metal. When $T = T_c$, N_c increases sharply, and for $T < T_c$ the condensed bosons determine the properties of the material.

The quantitative application of the theory has to be carefully handled. Introductory accounts are given by Ziman[4] and Rhoderick and Rose-Innes[5]. A more advanced account by Blatt[6] gives interesting historical details.

6

Fluctuations

We have briefly discussed the departures from the most probable values which are shown by the numbers of microparticles in various circumstances. In this chapter a more extensive theory will be presented. Historically, the theory of fluctuations is interesting because it made explicable several experimental phenomena which late-nineteenth-century physics, firmly adhering to thermostatics and classical mechanics, was either not able to explain or was able to explain only by *ad hoc* hypotheses. Examples of such phenomena are Brownian motion – the irregular motion of light macroparticles, e.g. pollen, suspended in a drop of liquid, which is readily observable with a microscope – and the spontaneous motion of light galvanometer suspensions, even when carefully shielded from air currents, etc. Many other examples occur in colloid science, sedimentation, etc., but perhaps the most important technically is electrical noise, which we discuss in detail later. Here, the basic phenomenon, which for technical reasons could hardly have been observed until well into the twentieth century, is that a direct current of electrons, which is very nearly constant at large values, i.e. when the number of electrons traversing the circuit in unit time is large, shows startling variations when the number of electrons is reduced until the flow is only a few electrons per unit time.

Thermodynamics cannot explain these fluctuations and, in fact, even denies their existence, because, as we shall see, a fluctuation into a less probable state of system involves a decrease of entropy, in contradiction to nineteenth-century ideas on the steady increase of entropy.

From what has just been said, the reader will appreciate that fluctuations, apart from their theoretical interest, are of basic importance in making accurate measurements of physical properties of a system and that fluctuations may impose limitations on measurement error which operate long before the limit due to quantum uncertainty is reached.

The problem of fluctuation theory is to relate the variations of the number density, energy, etc. to variations in the external parameters of the system under study.

6.1 Techniques of fluctuation theory

So far, we have established probability distributions for various physical situations and have taken the most probable value as identical with the mean value. This identity is, of course, not necessarily correct and, in fact, Khinchin's reformulation of statistical mechanical theory is largely concerned with the replacement of statements about the most probable value by statements about properly defined mean values and the subsequent use of the central-limit theorem of probability to yield accurate approximations to the true distribution function. However, from the viewpoint of the engineer and physicist, the important point about statistical mechanics is that its predictions agree with experiment and the fluctuations from the most probable or mean values need not be themselves calculated with very great accuracy. Even more, it is not necessary to characterize the distribution more exactly than can be done by prescribing the mean and the second moment. It should be emphasized that, until now, we have developed no apparatus for studying the *time* variation of any physical quantity. The question we shall ask is, if we have some physical quantity or quantities, say F_k and F_e, which are governed by the microcanonical distribution or by the canonical distribution, what is the probability of making a measurement of F_k or F_e and finding that it departs from the most probable value of F_k or F_e by a specified amount? The relevant second moments are

$$\langle (F_k - \langle F_k \rangle)^2 \rangle = \Delta^2 F_k, \quad \langle (F_k - \langle F_k \rangle) \cdot (F_e - \langle F_e \rangle) \rangle$$

and we shall often refer to the relative fluctuation $\Delta F_k / \langle F_k \rangle$.

Provided F_k and F_e depend only on the velocities or momenta of the particles, the second moments can be deduced directly from the distribution function, as we did for the energy and number density in Chapter 3. If F_k and F_e are functions of the coordinates, matters are more difficult and we need some further analysis. There are essentially two routes by which the necessary results can be obtained: we can either reduce the distribution function to a Gaussian shape, whose width specifies the second moment, or we can work directly from the canonical distribution, using, for example, some lemmas due to Gibbs. We shall start with the first technique.

6.2 Gaussian form of small fluctuations

First, consider systems in a microcanonical ensemble, i.e. with nearly fixed energy in ΔE at E. The entropy at equilibrium, i.e. when the system is in the most probable state, is S_0. The internal state of the system is assumed to depend on some parameter a which has an equilibrium value a_0. The entropy at value a is S_a. This means that measurements made on the system will average S_0 and a_0 but that, owing to fluctuations, deviations from these values will occur. Thus, if we make a sequence of measurements and use the time average, we shall expect to get S_0 and a_0 although the individual measurements may differ quite sharply from these values. We can, as yet, say nothing about the time variation, but the point being made is that the fluctuations are a consequence of molecular chaos and are not provoked by external factors. Since the entropy is a function of a, we can use the Boltzmann formula for entropy to write

$$dw = \text{const.} \times \exp[(S_a - S_0)/k]\, da \tag{6.1}$$

$$= \text{const.} \times \exp(\Delta S/k)\, da \tag{6.2}$$

together with $\int dw = 1.0$, which will fix the constant.

Since S_0 is the maximum entropy of the system, ΔS is negative and the probability of the fluctuation decreases sharply with increasing ΔS.

Next, we apply this result to the arrangement considered as the canonical ensemble, i.e. to a system in a heat bath at constant temperature T_0. The parameter a now specifies the state of the sub-system and, as a changes by $\mathrm{d}a$, the reservoir does work $\mathrm{d}W_a$ on the sub-system. The change of entropy is

$$\Delta S = \Delta S_0 + \Delta S_s \qquad 6.3$$

where ΔS_0 relates to the reservoir and ΔS_s to the sub-system. Equation 6.2 applies to the whole system, so

$$\mathrm{d}w = \text{const.} \times \exp[(\Delta S_0 + \Delta S_s)/k]\,\mathrm{d}a \qquad 6.4$$

But we can write

$$\Delta S_s = (\Delta E_s + p_0\,\Delta V_s - \mathrm{d}W_a)/T_0 \qquad 6.5$$

where p_0, T_0 are for the reservoir, i.e. practically for the whole lot, and E_s, V_s relate to the sub-system. Similarly,

$$\Delta S_0 = (\Delta E_0 + p_0\,\Delta V_0)/T_0 \qquad 6.6$$

Now, the total energy is conserved and so is the total volume. Therefore

$$\Delta E_s = -\,\Delta E_0$$
$$\Delta V_s = -\,\Delta V_0$$

therefore

$$\Delta S_s = -\,\Delta S_0 - \mathrm{d}W_a/T_0 \qquad 6.7$$

and equation 6.2 becomes

$$\mathrm{d}w = \text{const.} \times \exp(-\mathrm{d}W_a/kT_0) \qquad 6.8$$

We can also express this in another way, putting $\mathrm{d}W$ equal to the work done in a field with a potential U_a, i.e. $\mathrm{d}W = U_a - U_{a0}$. For small fluctuations we can expand U in a Taylor series:

$$U_a = U_{a0} + U'_{a0}(a - a_0) + \frac{U''}{2}(a - a_0)^2 + \ldots \text{etc.}$$

But, since a_0 represents an equilibrium condition, U_{a0} is a minimum of U_a and $U'_{a0} = 0$. Therefore

$$\mathrm{d}w \approx \frac{U''_{a0}}{2}(a - a_0)^2$$

and $$\mathrm{d}w = \text{const.} \times \exp\{-\,U''_{a0}(a - a_0)^2/2kT_0\}\,\mathrm{d}a \qquad 6.9$$

Now, the constant is clearly just

$$\text{const.} = \frac{1}{\int \exp\{-\,U''_{a0}(a - a_0)^2/2kT_0\}\,\mathrm{d}a} \qquad 6.10$$

so we have now found the distribution function required. The second moment of the fluctuations in a is now given, by definition, as

$$\Delta^2(a) = \langle (a - a_0)^2 \rangle$$

$$= \frac{\int (a - a_0)^2 \exp\{-U''_{a0}(a - a_0)^2/2kT_0\}\,da}{\int \exp\{-U''_{a0}(a - a_0)^2/2kT_0\}\,da} \qquad 6.11$$

where the integrals can be taken from $-\infty \rightarrow +\infty$ because their value is practically uninfluenced by large values of $(a - a_0)$. The integration is straightforward and we find

$$\Delta^2(a) = kT_0/U''_{a0} \qquad 6.12$$

Using this in equation 6.9, we find

$$dw = \frac{1}{(2\pi\Delta^2)^{1/2}} \exp\left\{\frac{-(a - a_0)^2}{2\Delta^2}\right\} da \qquad 6.13$$

which is the standard Gaussian distribution that we wished to find. The bigger Δ^2, the larger the fluctuation, since the exponential swamps the Δ^{-1} variation. Therefore, from equation 6.12, increasing the temperature T_0 increases the fluctuations, which seems sensible and certainly agrees with experiment for resistor noise. It is important to notice that, very frequently, one can use equation 6.12 directly to obtain the fluctuation one wishes to know.

Example 1. To find the volumetric fluctuations in an ideal gas at constant pressure

The free energy F is a potential for the variables V and p, and, from Appendix 2,

$$p = -(\partial F/\partial V)_T \qquad \therefore \partial^2 F/\partial V^2 = -\partial p/\partial V$$

Therefore, from equation 6.12,

$$\Delta^2 V = \langle (V - V_0)^2 \rangle = -kT_0\,\partial V_0/\partial p$$

But $p_0 V_0 = NkT_0$

therefore $\Delta^2 V = V_0{}^2/N$

or $\Delta V/V_0 = 1/\sqrt{N}$ 6.14

Alternatively, $\Delta T/T_0 = 1/\sqrt{N}$ 6.15

The last equation relates to a real instrument: the constant-pressure gas thermometer. If $N = 10^6$, the error is 0.1%; but there are over 6×10^{26} (Avogadro's number) molecules in a kilomole of gas, thus the mass of gas could be reduced by a factor of $O(10^{-21})$ before even a 0.1% error would be reached.

Example 2. Fluctuations in a torsion galvanometer

Here, let the suspended mirror be in equilibrium at an angle $\theta = 0$. For small angular rotations, the potential energy is $\frac{1}{2}\alpha\theta^2$ where $\alpha = (\pi r)^2 G/2l$, G = torsion modulus of the wire material, and the wire is length l, radius r.

Then $U'' = \alpha$

and $\Delta^2\theta = kT_0/\alpha$ 6.16

Equation 6.16 means that the galvanometer mirror will spontaneously rotate, on the average, between the angles $+\Delta\theta$ and $-\Delta\theta$ or $\pm\sqrt{(kT_0/\alpha)}$, with larger excursions being relatively improbable, following equation 5.13.

Thus, for a *single* measurement of a current, the minimum detectable current would be that deflecting the mirror by $\Delta\theta_0 = \sqrt{(kT_0/\alpha)}$. However, if a long sequence of measurements could be made, smaller currents than this could be measured, because the mean θ would have shifted from zero. In other words, if the deflection were measured as a time average over a long time, current changes smaller than the 'noise' could be measured. We shall return to this question in a much more thorough way later in the book.

The reader will probably have realized that these examples could have been solved using the equipartition theorem. The advantage of the present treatment is that we have equation 6.13, which tells us the relative frequency of fluctuations of n times the second moment, which clearly fall off rapidly for $n > 1$.

6.3 Derivation of the Gaussian distribution from measurements

In section 1.6, we showed that the characteristic function is related to the cumulants of the distribution function by

$$C(s) = \exp\left(\sum_{n=1}^{\infty} \frac{(js)^n}{n!} k_n\right)$$

and that the first two cumulants, k_1 and k_2, are respectively the mean and the dispersion of the distribution function. Suppose that we have measured k_1 and k_2 for a particular distribution function, i.e. suppose we have measured the mean and deviation of some physical quantity. Then we can write

$$C(s) \approx \exp(jsk_1 - s^2k_2/2) \tag{6.17}$$

and, inverting this result by the Fourier integral theorem,

$$f(x) \approx \frac{1}{2\pi}\int_{-\infty}^{+\infty} ds \,.\, \exp(-jsx)\exp(jsk_1 - s^2k_2/2)$$

$$\approx \frac{1}{(2\pi k_2)^{1/2}} \exp\left\{-\frac{(x-k_1)^2}{2k_2}\right\} \tag{6.18}$$

This result is just a Gaussian distribution and, by adjusting the notation, can be made to coincide with equation 6.13. Here we have shown that a Gaussian distribution is the best representation we can obtain from a knowledge of two cumulants only. This (a) accounts for the fundamental importance of the Gaussian form in statistical mechanics and (b) very much reduces the degree of arbitrariness in the derivation of equation 6.13.

From another viewpoint this is an application of the central-limit theorem of probability theory, which, very roughly, states that, whatever the form of $f(x)$, if the events x are statistically independent and $f(x)$ tends rapidly to zero as $x \to \infty$, the distribution tends to the Gaussian form for a very large number of events. Khinchin[1] uses this theorem as a basis for his reformulation of statistical mechanics, and this work includes a rigorous proof of the central-limit theorem and careful estimates of the error when moments of up to the fifth order exist. The simplest of these estimates is that equation 6.18 has an error of order $(1/N^{3/2})(1 + |x - k_2|)$, which is certainly negligible in most of the problems we have discussed.

Here, we have discussed a one-dimensional Gaussian form. The method can be extended to two or many more dimensions and is extremely powerful. We shall return to it later on.

6.4 The approach to equilibrium

We can now draw some conclusions about the approach to equilibrium, as it appears from the viewpoint of statistical mechanics. First, we have seen that if we observe some property of the sub-system, in equilibrium, instead of the property remaining fixed as predicted by classical thermostatics, we shall observe fluctuations in it. These take place in the positive and negative directions, but their long-time average is zero. However, the second moment is finite.

Now consider what happens as the sub-system approaches equilibrium. As a concrete example, imagine a very small drop of dye introduced into a water film on a microscope slide. How will the dye spread out from the point of introduction? We can, as yet, say little about the early stages of this process, but in the later stages, when the dye distribution is *nearly* uniform, we can reason that the departure from equilibrium is small enough for the theory developed above to apply, except that there are fluctuations corresponding with a *gain* in entropy. These are much more probable, according to the basic expression, equation 6.2, than fluctuations in which the entropy decreases. Thus, the fluctuations will, on average, carry the system towards the equilibrium state in which the entropy is maximized.

Although it is simple to understand the manner in which fluctuations bring the system back to equilibrium after a small displacement, we still wish to know two more important things: (a) what factors govern the behaviour of the fluctuations in time? and (b) how does the system, starting from initial conditions, which in the general case are not in the least like the equilibrium state, *first* reach equilibrium? It must be immediately stated that, at present, a fully satisfactory treatment of (b) is not available. Later, some kinetic theories which provide a partial answer will be discussed.

Returning to (a), we imagine that a system in equilibrium with a heat bath is made the subject of a time sequence of measurements, or even a continuous sequence of measurements; for example a very small and highly sensitive pressure sensor

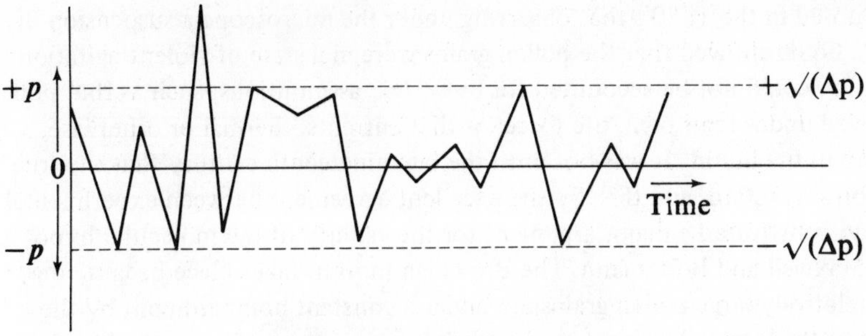

Fig. 6.1 A very rough attempt to illustrate a random motion.

might be included in the boundary of the system and the amplified readings displayed on a chart recorder. From the consideration given in earlier sections we can derive the mean reading, and from section 6.2 we can calculate the deviation. Suppose that our measurement instrumentation is arranged to discard the mean pressure reading, then the chart will appear roughly as in Fig. 6.1 (it is extremely difficult to sketch random waveforms in a convincing way). The pressure very rarely deviates *outside* the limits $\pm (\Delta p)^{1/2}$, but the variation is random in time.

We should very much like to know what processes control the frequencies included in the waveform $p(t)$. Obviously, the measurement system imposes a bandwidth limit, but we may get round this by several obvious techniques – by using a very broadband system or by using different systems to explore several frequency ranges – so this difficulty is inessential. Moreover, experiment shows that for very many fluctuating processes the generalized 'power'* per unit bandwidth is constant or, in other words, if we measure the power over a time T centred on t_1, it turns out to be identical with that in T centred on t_0, if only T is sufficiently long. And so on, for $t_0, \ldots, t_n$. This property defines a '*stationary random process*'. But statements about the power still tell us nothing about the frequency decomposition, which is obviously vital if we wish to design techniques for improving the accuracy of measurement, for example. To get information about the fundamental time scale, we must introduce new physics into the problem. This will be done by a more detailed examination of the collision process or processes, which randomize the particle motions.

To conclude, the reader will have realized that the assumption that the long-time average and the ensemble average of a fluctuating process are indentical is implicit in this discussion. The assumption can be regarded as proven by a suitably weak ergodic hypothesis or by comparison of the theory with experiment.

6.5 Brownian motion

For historical reasons and because of the fundamental importance of the ideas and methods involved, we now discuss the theory of Brownian motion. Robert Brown, a

*In the example used, the pressure has been converted to a voltage or current driving a recorder of some sort. Assuming linear transducers, it is immaterial whether we discuss the input or the output conditions. Thus, the term 'power' is appropriate.

botanist, noted in the 1820's that observing under the microscope a suspension of pollen in a liquid showed that the pollen grains were in a state of violent agitation. The agitation could not be accounted for by *ad hoc* assumptions, such as that pollen grains moved under their own 'life force' or that currents, thermal or otherwise, were set up in the liquid. It was not until the late nineteenth century that the true explanation was found, and this, by the excellent agreement between experimental and theory, constituted a major argument for the validity of the molecular hypotheses of Maxwell and Boltzmann. The Brownian motion takes place because the light but relatively large pollen grains are under a constant bombardment by the molecules of the liquid. These molecules exchange energy with the grain, but the direction of the velocity acquired by the grain is purely random. Additionally, we must recognize that the movement of a large light body is resisted by a frictional force of macroscopic nature and which can be assessed by Stoke's law. The frictional force acts continuously on the body whereas the period of an individual collision is short. We can, then, already isolate two characteristic times, viz. the duration of a collision and the free time between collisions.

The Brownian motion was first adequately explained by Einstein,[2] in a series of papers published between 1905 and 1908. Here, we introduce the topic using a formulation due to Langevin, which makes the particulate dynamics extremely simple. Langevin[3] focuses attention on the motion of the large particle of mass M, whose centre of mass is at $x(t)$ and whose velocity is $v = x$. — $\dot{x}$?

He divides the force into the sum of two parts, one of which is the slowly varying macroscopic force and the other the very rapidly varying random force, whose time average is zero but which has a root-mean-square value which is finite. Then, for one-dimensional motion,

$$M\ddot{x} + \beta\dot{x} = F + F_1(t) \qquad 6.19$$

where β = frictional coefficient, F_1 = random or stochastic force.

Equation 6.19 is, of course, clear and obvious from the viewpoint of macroscopic dynamics. The question of its validity in the microscopic behaviour of particles is more subtle. In our Gibbsian statistical mechanics we have made the basic assumption that all the processes are reversible and that a particle would reverse accurately the forward trajectory, if time were reversed. This assumption *appears* to be backed up by the detailed classical consideration of binary collisions, in which energy is conserved. But, even in a tenuous real gas the binary collisions do not take place in isolation and the colliding particles are really interacting, however weakly, with all the other particles of the system. Thus, even in a very simple case, classical collision theory neglects a perturbation term. This term will certainly mean that precise reversibility is lost, and in an improved theory it is, at least, reasonable to introduce frictional or dissipation effects. We shall return to this topic later; here the point is to realize the important physical content of the Langevin equation.*

To proceed with the solution of the Langevin equation, one very simple argument

*Many writers dismiss Langevin's work with disparaging comments about the elementary nature of the analysis. This seems to me unfair. In my view, Langevin's theory is one of the seminal ideas of modern kinetic theory.

is the following. First, consider the motion *over a short time*. In this case, it is permissible to ignore the steady force F, since the integrated effect will be negligible. Then, multiply equation 6.19 by x to get

$$M x\ddot{x} = M[\mathrm{d}(x\dot{x})/\mathrm{d}t - \dot{x}^2] = -\beta x\dot{x} + xF_1(t) \qquad 6.20$$

Then, average equation 6.20 over an ensemble of particles. x and F_1 are independent, therefore $\langle xF_1\rangle = \langle x\rangle . \langle F_1\rangle = 0$, since $\langle F_1\rangle = 0$, and we can put $\dot{x}^2 = kT/M$ from classical equipartition theory.

Then $M\mathrm{d}/\mathrm{d}t(x\dot{x}) = kT - \beta\langle x\dot{x}\rangle$ 6.21

or $\langle x\dot{x}\rangle = C\exp(-\beta t/M) + kT/\beta$ 6.22

If we now impose the condition on the ensemble that each particle starts from $(x, t) = (0, 0)$, $C = -kT/\beta$, we obtain

$$\tfrac{1}{2}\frac{\mathrm{d}\langle x^2\rangle}{\mathrm{d}t} = \langle x\dot{x}\rangle = \frac{kT}{\beta}\left\{1 - \exp\left(-\frac{\beta t}{M}\right)\right\} \qquad 6.23$$

or $$\langle x^2\rangle = \frac{2kT}{\beta}\left[t - \frac{M}{\beta}\left\{1 - \exp\left(-\frac{\beta t}{M}\right)\right\}\right] \qquad 6.24$$

Equation 6.24 gives the mean square deviation of a particle which started from (0, 0) after time t. Initially, equation 6.24 shows that $\langle x^2\rangle \approx (kT/M)t^2$, and this in turn means that, for very short times $t < M/\beta$, the particle behaves as though it were moving with a constant velocity, equal to the thermal velocity. For $t \to \infty$, $\langle x^2\rangle \to (2kT/\beta)t$ and the frictional force has become dominant.

To make use of these results, Stoke's law is used to give β. For spherical particles radius a in a fluid of viscosity η,

$$\beta = 6\pi\eta a \qquad 6.25$$

Before Langevin's work, Einstein had shown that, for long times, the motion of the Brownian particle could be described by the diffusion equation

$$\partial\rho/\partial t = D\,\partial^2\rho/\partial x^2 \qquad 6.26$$

where ρ is the particle density. Solving equation 6.26, we get

$$\langle x^2\rangle = 2Dt \qquad 6.27$$

so that we must identify the diffusion coefficient D with $2kT/\beta$ if the two theories are to agree.

Then $D = kT/\beta = kT/6\pi\eta a$ 6.28

so that we have achieved a relationship between the diffusion coefficient, known parameters η and a of the system, and the Boltzmann constant.* In the 1920's.

*The definition of D leads also to the 'Einstein relation' which is widely used in semiconductor physics. Consider a charged particle in an electric field so that $F = eE$ and disregard the stochastic force. The final velocity of the particle is $eE/\beta = v$ and the mobility is $\mu = e/\beta$. Using equation 6.28, $\mu = eD/kT$ or, more generally, if the mobility is defined as the terminal velocity for unit force, $D = \mu kT$ for *any* particle. This is called the 'Einstein relation'.

Perrin used equation 6.28 to establish values of k which were more accurate than the then accepted value.

We return now to equation 6.24 and put $M/\beta = \tau$, clearly the time constant of the system. We have already discussed the behaviour of $\langle x^2 \rangle$ for short and long times, and closer inspection of equation 6.24 shows that this means we can categorize times as much shorter than τ or as much longer than τ. This quantity arises from the analysis of the system and is the first example we have encountered of a class of times which are called 'relaxation times'. These are characteristic times which govern the evolution of the system from one state to another, in this case from the state in which only the initial stochastic effects are important, since we assumed the initial stochastic velocity was $\langle v^2 \rangle = kT/M$, to a state in which the collective behaviour of the particles as expressed by the frictional coefficient is the dominant feature. It turns out that appropriate relaxation times can be defined for many different types of system, for example for electron flows in semiconductors and charged-particle flows in plasmas. We shall shortly see how the relaxation time is taken into non-equilibrium statistical mechanics in a particular example of the Boltzmann transport equation.

6.6 The time correlation function or autocorrelation function

We may visualize the effect of each collision on the pollen grain as that of an impulse, very short in duration, which imparts a velocity to the particle. The friction damping destroys this velocity so that, a long time after the impulse, the particle is no longer influenced by the impulse. Thus, the response to each impulse

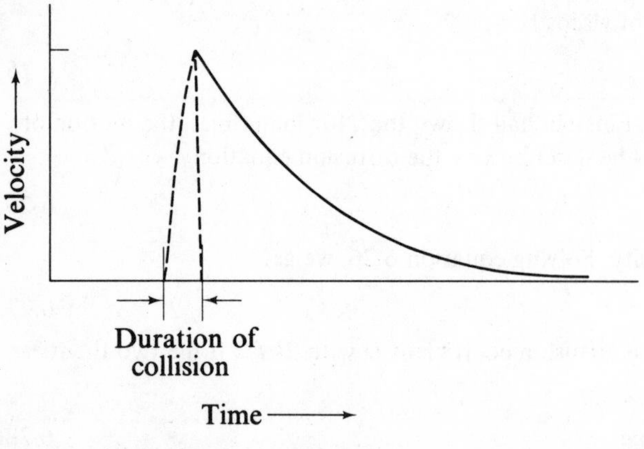

Fig. 6.2 The effects of a collision.

is damped as indicated in Fig. 6.2. The velocity of the particle is then due to a series of such impulses, occurring at random instants of time, and therefore constitutes what is nowadays called a 'stochastic' process.

Fig. 6.3 illustrates a chain of impulses, shown for convenience as of equal amplitude, and the same train with each pulse displaced from t_n to $t_{n+\tau}$ where the

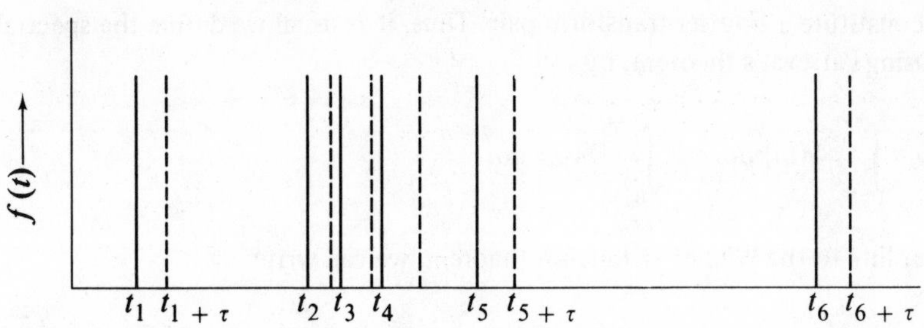

Fig. 6.3 Formation of the autocorrelation of a chain of impulses.

t_n are purely random instants and τ* is a known constant. The time correlation function, or 'autocorrelation function' (a.c.f.) as it is called in communications engineering, is defined by

$$K(\tau) = \langle f(t) \,.\, f(t+\tau)\rangle \qquad 6.29$$

or, more precisely, by

$$K(\tau) = \lim_{T\to\infty} \frac{1}{T}\int_0^T f(t) \,.\, f(t+\tau)\,\mathrm{d}t \qquad 6.30$$

τ can be either positive or negative and we shall soon see that $K(\tau)$ is a symmetric function of τ. We emphasize again that time averaging and ensemble averaging are to be taken as equivalent processes. By definition, we see that $K(0) = \langle f^2(t)\rangle$ and is therefore the mean-square value of the stochastic process. To prove the symmetry property, we write $t = t_1 - \tau$ in equation 6.29; then

$$\langle f(t) \,.\, f(t+\tau)\rangle = \langle f(t_1-\tau) \,.\, f(t_1)\rangle$$
$$= \langle f(t_1) \,.\, f(t_1-\tau)\rangle$$
$$= K(-\tau)$$

$$\text{Thus} \quad K(\tau) = K(-\tau) \qquad 6.31$$

which was to be proved.

The last of the simple properties is that $K(\infty) \to 0$, which may be taken as stating that no real system can possibly remember an impulse given to it a very very long time ago.

The a.c.f. has less simple properties, which are summed up in the 'Wiener–Khinchin' theorem, the proof of which is given in Appendix 4. Stated in words, the theorem is that the a.c.f. and the spectral density function of the same set of

*The use of the symbol τ here might be confusing since it is *not* the relaxation time but an arbitrary time interval. However, it is so extensively used in the modern literature that it seems necessary to adopt it.

impulses constitute a Fourier-transform pair. Thus, if as usual we define the spectral density, using Parseval's theorem, by

$$\langle f^2(t)\rangle = \int_{-\infty}^{+\infty} \Phi(\omega)\,\mathrm{d}\omega = 2\int_0^{+\infty} \Phi(\omega)\,\mathrm{d}\omega$$

then, according to the Wiener–Khinchin theorem, we can write

$$\text{a.c.f.} = K(\tau) = \int_{-\infty}^{+\infty} \mathrm{d}\omega\,.\,\Phi(\omega)\exp(\mathrm{j}\omega\tau) \qquad 6.32$$

$$\text{and} \quad \Phi(\omega) = \frac{1}{2\pi}\int_{-\infty}^{+\infty} \mathrm{d}\tau\,.\,K(\tau)\exp(-\mathrm{j}\omega\tau) \qquad 6.33$$

Put into electrical terms, $\Phi(\omega)$ describes the spectral distribution, i.e. the distribution in frequency, of the power in the random waveform $f(t)$ or $v(t)$ or $i(t)$ and, for example, is just a constant times the bandwidth for shot noise or resistor noise up to very high frequencies. We also remark that, since $K(\tau)$ is real and symmetrical, we can simplify equations 6.32 and 6.33 to their real forms:

$$K(\tau) = \int_{-\infty}^{+\infty} \mathrm{d}\omega\,\Phi(\omega)\cos\omega\tau \qquad 6.34$$

$$\Phi(\omega) = \frac{1}{2\pi}\int_{-\infty}^{+\infty} \mathrm{d}\tau\,K(\tau)\cos\omega\tau \qquad 6.35$$

We have said earlier that stationary random functions have spectral power densities which are independent of the time of observation, therefore the Wiener–Khinchin relationships prove that the a.c.f. is the same for all members of an ensemble of such functions and for all representatives of such functions.

To summarize, if we know or can calculate the spectral density function, we can find the a.c.f. from equation 6.34. Alternatively, if we know $K(\tau)$ we can find the spectral density from equation 6.35, and, depending on the information given in specific problems, it may be easier to use one relationship rather than the other. Either function contains most of the relevant information on the stochastic process.

As an example of the use of these results, let us revert to Brownian motion and derive the a.c.f. of the particle velocity and the spectral distribution of the same variable. If we rewrite the stochastic part of equation 6.19 in terms of velocity, we get

$$M\dot{u} + \beta u = F_1(t)$$

$$\text{or} \quad u(t_1) = u_0\exp(-\beta t_1/M) + \exp(-\beta t_1/M)\int_0^{t_1} \exp\beta\zeta\,F_1(\zeta)\,\mathrm{d}\zeta \qquad 6.36$$

where u_0 is the velocity immediately after the collision. Taking the ensemble average for particles with the same value of u_0, we find

$$u(t_1) = u_0 \exp(-\beta t_1/M) \tag{6.37}$$

because $\langle F_1(t)\rangle = 0$, by definition. Thus, the velocity falls off exponentially and we can also see that

$$u(t_1 + \tau) = u_0 \exp\{-\beta(t_1 + \tau)/M\} = u_0 \exp(-\beta t_1)/M \exp(-\beta\tau/M)$$

Then, using equations 6.30 and 6.31, we find directly that

$$K_u(\tau) = u_0^2 \exp(-\beta|\tau|/M) \tag{6.38}$$

and, by Fourier inversion,

$$\Phi_u(\omega) = \frac{u_0^2}{\pi} \frac{\beta/M}{(\beta/M)^2 + \omega^2} \tag{6.39}$$

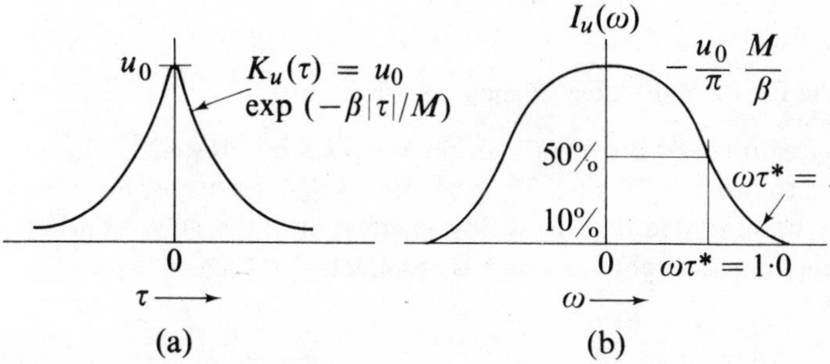

Fig. 6.4 The autocorrelation function for a Brownian collision and the corresponding spectral distribution function.

These functions are illustrated in Fig. 6.4. Equation 6.39 is a 'resonance' curve, i.e. the response curve of a resonant circuit plotted about the resonant frequency.

To discuss these results, we first note that if we put in the relaxation time, which we now call τ^*,

$$K_u(\tau) \propto \exp(-|\tau|/\tau^*) \tag{6.40}$$

and so has fallen to about 40% of the initial value after one relaxation time. Therefore, the a.c.f. is numerically near its maximum value for times considerably shorter than the relaxation time and nearly zero for times much longer than the relaxation time, in perfect agreement with our earlier limiting cases in which stochastic variations or, more strictly, their initial values were important for short times while the macroscopic frictional force took over at long times.

We can rewrite equation 6.39 in the form

$$\Phi_u(\omega) = \Phi_u(0) \frac{1}{1 + (\omega\tau^*)^2} \tag{6.41}$$

so that the spectral density is halved at the frequency which makes $\omega\tau^* = 1.0$ and is only 10% of the initial value when $\omega\tau^* = 3.0$. In principle, a measurement of $\Phi(\omega)$, using narrow-band instrumentation giving $\Phi(\omega_1)$ in the small range $\Delta\omega$ at ω_1 for a range of ω_1's, would yield a value of τ^*. The difficulty is that, in those situations in which one would like to do this, τ^* tends to be extremely short, in a gas $0(10^{-10})$s, so that present-day instrumentation tends to limit one to the relatively uninformative region $\omega\tau^* < 1.0$. It is appropriate to say now that the a.c.f. $\exp[-(\text{const.})|\tau|]$ is met with in many fields other than that of Brownian motion, which is why we have discussed it in considerable detail.

The results of this section have provided us with all the information that we can hope to get about both the ensemble and time behaviour of the fluctuations about the mean of physical quantities. We have, so far, considered only a few particular examples, but later on we shall discuss the practically very important question of electrical noise in full detail and from an even more general viewpoint. For the moment we can leave particular problems and take up the discussion of a different and very useful technique for establishing the basic probabilistic relationships.

6.7 Diffusion and the Einstein–Fokker–Planck equation

The first successful treatment of Brownian motion was given by Einstein[2], who reduced the problem to one of diffusion. To speculate on the reasons for his using this approach, it can be suggested that the diffusion equation is the simplest differential equation in mathematical physics which is asymmetric in time. The one-dimensional form is

$$\partial f/\partial t = D\, \partial^2 f/\partial x^2 \qquad 6.42$$

so that the solutions for positive and negative t are, in general, not identical. Since the Brownian motion is irreversible, it was important to include this feature in the analysis. Another indicator for this approach may have been that, if one considers introducing a small drop of moist pollen into a thin film of liquid on a slide, it is good sense to consider that the equilibrium state is established by diffusion.

Einstein's argument (reference 2, p. 12 *et seq.*) can be summarized as follows. Assume that each single grain executes a motion which is independent of that of all other particles. The relaxation time τ is small compared with the time of observation but is defined as long enough for the motions of a particle in two consecutive intervals of τ to be independent.

Let the number of suspended grains be N and let the x coordinate change by Δ in one relaxation time. Δ is a stochastic variable, symmetric about $\Delta = 0$. Then, the number of particles $\mathrm{d}N$ which are displaced by $\mathrm{d}\Delta$ at Δ is

$$\mathrm{d}N = Np(\Delta)\,\mathrm{d}\Delta \qquad 6.43$$

where $p(\Delta)$ is the probability distribution and

$$\int_{-\infty}^{+\infty} p(\Delta)\,\mathrm{d}\Delta = 1 \qquad 6.44$$

$p(\Delta)$ is sharply peaked near $\Delta = 0$, since only small displacements are possible.

Let $\rho(x, t)$ be the density of particles at (x, t). The number of particles in the volume enclosed by the planes x and $x + dx$ at time $t + \tau$ is

$$\rho(x, t+\tau)\,dx = dx \int_{-\infty}^{+\infty} \rho(x+\Delta)\,p(\Delta)\,d\Delta \qquad 6.45$$

where the infinite limits are allowed because of the assumed form of $p(\Delta)$.

Then, in view of the smallness of τ,

$$\rho(x, t+\tau) \approx \rho(x, t) + \tau\, \partial\rho/\partial t$$

Also, $p(x + \Delta, t) = \rho(x, t) + \Delta \dfrac{\partial \rho}{\partial x} + \dfrac{\Delta^2}{2!}\dfrac{\partial^2 \rho}{\partial x^2}$

$$\therefore \rho(x, t) + \frac{\tau \partial \rho}{\partial t} = \rho(x, t)\int p(\Delta)\,d\Delta + \frac{\partial \rho}{\partial x}\int \Delta p(\Delta)\,d\Delta + \frac{\partial^2 \rho}{\partial x^2}\int \frac{\Delta^2}{2!}\,p(\Delta)\,d\Delta$$

$$= \rho(x, t) + \frac{\partial^2 \rho}{\partial x^2}\int \frac{\Delta^2}{2!}\,p(\Delta)\,d\Delta$$

because of equation 6.44 and the fact that terms odd in Δ vanish on integration.

$$\therefore \frac{\partial \rho}{\partial t} = \frac{1}{\tau}\int \frac{\Delta^2}{2!}\,p(\Delta)\,d\Delta\;\frac{\partial^2 \rho}{\partial x^2}$$

$$= D\frac{\partial^2 \rho}{\partial x^2} \qquad 6.46$$

$$\text{if}\quad D = \frac{1}{\tau}\int_{-\infty}^{+\infty} \frac{\Delta^2}{2}\,p(\Delta)\,d\Delta \qquad 6.47$$

$$= \frac{1}{2\tau}\langle \Delta^2 \rangle \qquad 6.48$$

where the last line follows from the definition of $p(\Delta)$.

We have already used these last results in equation 6.26 *et seq.*, but we can now go further. The solution of equation 6.46, which is valid for $\rho(x, t) = \rho(x_0, t_0)$ and for $\int \rho(x, t)\,dx = N$ is

$$\rho(x, t) = \frac{N}{\sqrt{\{4\pi D(t - t_0)\}}}\exp\left\{\frac{-(x - x_0)^2}{4D(t - t_0)}\right\} \qquad 6.49$$

a Gaussian form, but one which both space and time coordinates appear.

The Einstein treatment gives the physically important result of the value of the diffusion coefficient and evolution of the density as described by equation 6.49. It does not, in the present form, yield the general result equation 6.24 for the mean-square value of x. To do this, equation 6.46 must be generalized. The physical basis for the generalization is that the assumption of equal probability for positive and negative displacements must be abandoned. Clearly, if the motion takes place in a field of force, displacement in the sense determined by the force is more probable than in the opposite sense. Therefore, in the more general case, the term $(\partial\rho/\partial x)\int\Delta\, p(\Delta)\mathrm{d}\Delta \neq 0$ and must be retained. If this is done, we get

$$\frac{\partial\rho}{\partial t}=\frac{1}{\tau}\;\Delta\;\frac{\partial\rho}{\partial x}+D\frac{\partial^2\rho}{\partial x^2} \qquad 6.50$$

or, more strictly,

$$\frac{\partial\rho}{\partial t}=\frac{1}{\tau}\left\langle\Delta\frac{\partial\rho}{\partial x}\right\rangle+\frac{1}{2\tau}\left\langle\Delta^2\frac{\partial^2\rho}{\partial x^2}\right\rangle \qquad 6.51$$

These equations are one-dimensional examples of what are called 'Einstein–Fokker–Planck' or, more commonly, 'Fokker–Planck' equations.

The Fokker–Planck equation is important in kinetic theory, and we shall wish to use it frequently; therefore it is worth quoting in a more general context, the derivation being relegated to Appendix 5. Let $\rho(\lambda, t)$ be the probability that a system is in state $\mathrm{d}\lambda$ at λ. Under random influences, the state of the system changes and, for example, p changes from $p(\lambda_0, t_0)$ to $\rho(\lambda, t_0+\Delta\tau)$ in time Δt. The Fokker–Planck equation depends on the following important assumption, viz. that the transition probability, that is the probability of the system making the transition specified, depends only on λ_0, λ and Δt; moreoever, the transition probability is the conditional probability that, it being specified as certain that the system is at state λ_0, t_0, the system changes to state λ in Δt. Thus, the system has no 'memory' for what happened to it in reaching the state λ_0. Such processes are called 'Markov' processes in probability theory, and it is by no means always clear that a particular physical process is exactly Markovian. However, here we assume that our process is, in fact, Markovian. The transition probability is $w(\lambda_0, \lambda, t)$ and it is normalized to unity, as is $p(\lambda, t)$. Lastly, it is assumed that w is a rapidly decreasing function of $|\lambda-\lambda_0|$, so that most changes of state are small. Then we can write an integral equation, which is called 'Smoluchovski's equation', relating the probabilities. This is

$$p(\lambda, t_0+\Delta\tau)\,\mathrm{d}\lambda=\mathrm{d}\lambda\int\rho(\lambda_0, t_0)\,w(\lambda_0, \lambda, \Delta t)\,\mathrm{d}\lambda_0 \qquad 6.52$$

where the integral is taken over all possible λ.

By processes very similar to those used by Einstein in deriving equation 6.46, we can show (Appendix 5) that

$$\frac{\partial p}{\partial t}=-\frac{\partial}{\partial\lambda}(ap)+\frac{\partial^2}{\partial\lambda^2}(Dp) \qquad 6.53$$

where

$$a(\lambda_0) = \lim_{\Delta t \to 0} \int \frac{\lambda - \lambda_0}{\Delta t} w(\lambda_0, \lambda, \Delta t)\, d\lambda \qquad 6.54$$

$$D(\lambda_0) = \lim_{\Delta t \to 0} \int \frac{(\lambda - \lambda_0)^2}{\Delta t} w(\lambda_0, \lambda, \Delta t)\, d\lambda \qquad 6.55$$

If we let $G = ap - \frac{\partial}{\partial \lambda}(Dp)$ 6.56

equation 6.53 becomes

$$\frac{\partial p}{\partial t} = \frac{\partial G}{\partial \lambda} \qquad 6.57$$

Equation 6.57 has the form of an equation for continuity, and we can therefore call G the 'probability *current*'* or, more intuitively, G represents the flow of representative points in a phase space.

We consider now the application of equation 6.53 to the case of Brownian motion. We wish to find $p(x, t)$, the probability of finding the particle at (x, t), starting from (0, 0). Equation 6.54 shows that a is $\lim_{\Delta t \to 0} \langle x/\Delta t \rangle$, the average velocity of the particle. From equation 6.55, D is the mean square displacement as before.

Then $\frac{\partial p}{\partial t} = -\langle v \rangle \frac{\partial \rho}{\partial x} + D \frac{\partial^2 \rho}{\partial x^2}$

whose solution, for the given initial and final conditions, is

$$\rho(x, t) = \frac{1}{\sqrt{(4\pi D t)}} \exp\left\{ -\frac{(x + \langle v \rangle t)^2}{4Dt} \right\} \qquad 6.58$$

This result compares with Einstein's result, equation 6.49, in which $\langle v \rangle = 0$.

At this point the reader may appreciate some remarks on methods for solution of the Fokker–Planck equation. Provided a and D are constants independent of coordinates and time, there is no particular difficulty. Perhaps the most powerful technique is to use a Fourier transformation in time to reduce the equation to an ordinary differential equation which can be solved by the Laplace transform. Many examples of the application of this technique, applied to generalized diffusion problems, are given by Carslaw and Jaeger[4]. Separation of variables sometimes works and examples are given by Stratonovich[5]. However, in less simple applications, in anisotropic media particularly, when a and D have to be replaced by tensors, the difficulties of solution are very much greater and we shall discuss these topics when we consider plasma theories.

*Compare with the electrical continuity equation for charge density ρ, charge velocity u. $\partial\rho/\partial t = \partial(\rho u)/\partial x = \partial J/\partial x$, where J = current density.

7

Kinetic Theories

Statistical mechanics, of its nature, deals with processes in either microcanonical or canonical ensembles, which by definition represent systems in thermodynamical equilibrium. Therefore, changes are fundamentally reversible. In Chapter 6 we discussed Langevin's idea of introducing a frictional force to give irreversible processes, but this is a purely phenomenological device, i.e. the friction is introduced into the theory from outside and does not arise as a consequence of the theory itself. We now wish to study irreversible processes from a more basic viewpoint. This study is called 'kinetic theory' and it is of the greatest possible importance today, because of its bearing on plasma theory and the instabilities of plasma discharges which have, so far, prevented us from realizing nuclear fusion on a practical scale. From the theoretical side, we may also expect that kinetic theory will produce quantities similar to the friction in their effects, but which will arise naturally from the theory itself. This hope is only partially realized as yet, in spite of the enormous effort put into kinetic theory in the last thirty years.

Physical kinetics seeks to study the evolution of the distribution function of any physical variable in time. If one can calculate the transition probability from one microstate to another, one can calculate the rate of change of the distribution function or, finally, the new non-equilibrium distribution function. The equations governing the change of the distribution function from equilibrium to non-equilibrium are called 'kinetic' equations. It should be made clear that the more general the kinetic equation, the more difficult it is to solve. For this reason, present-day kinetic theory is in the state where a large variety of approximate theories, which are valid only under specified conditions, have been derived and solved. Often, the solutions are themselves only approximate, so that we are dealing with approximations to approximations. However, progress is always being made in the direction of generalization and improvement so it is not too much to hope that a general theory will, eventually, emerge.

7.1 The Boltzmann transport equation

The first serious attempt at a kinetic theory was made by Boltzmann, starting from a study of the time evolution of the Maxwell velocity distribution function. The ideas therefore are basically concerned with the behaviour of gases which are not too dense. The theory also applies to, for instance, electrons in semiconductors, but in following the derivations it is useful to keep a gas in mind.

There are many ways of deriving the Boltzmann equation, and the validity of the equation has been extensively discussed* e.g. by Grad[1,2]. These discussions are

*Historically, the intense controversy caused by the publication of the equation and the criticisms of it by Loschmidt, Zermelo and others, who could not accept the molecular nature of matter, are covered by the collection of papers edited by Brush[3].

more interesting to mathematicians than to engineers but it must also be said that, in those cases where the equation can be solved, the results agree with experiment and with other theories sufficiently well for us to believe that it is essentially correct, in the range where it can be applied. For these reasons, we present a very simple and intuitive derivation.

Consider a gas in which the molecules are subject to the action of a force $\boldsymbol{F}$, which is independent of the particulate velocity but which may be a function of the space coordinates. The gas is so tenuous that the time spent by a molecule in collisions is a small fraction of the total lifetime or, more concretely, the collision time is much less than the mean free time. If we then consider a small element of physical volume in the gas, the number of molecules in the volume element will change for two different reasons:

1. Since the molecules are all in a state of motion under the action of the force, there will be a flow of particles into and out of the volume element. This flow can be calculated, just as in hydrodynamics, by Eulerian methods and, in fact, just depends on the Eulerian derivative of the distribution function.

2. The collisions will scatter out of the volume element some particles which, had the collision not taken place, would have remained within it. Conversely, some particles will be scattered into the volume element which would not otherwise have entered it. This term, then, depends on the actual physics of the collision process.

First, we compute (1). Let $\mathrm{d}n$ = number of molecules with representative points in $\mathrm{d}\mu$ at time t. We have

$$\mathrm{d}\mu = \mathrm{d}q_1\,\mathrm{d}q_2\,\mathrm{d}q_3\,\mathrm{d}p_1\,\mathrm{d}p_2\,\mathrm{d}p_3 \tag{7.1}$$

and $$\mathrm{d}n = f(\boldsymbol{r}, \boldsymbol{p}, t) \tag{7.2}$$

The total time derivative of f is, in coordinate form,

$$\frac{\mathrm{d}f}{\mathrm{d}t} = \frac{\partial f}{\partial t} + \frac{\partial f}{\partial q_1}\cdot\frac{\partial q_1}{\partial t} + \frac{\partial f}{\partial q_2}\cdot\frac{\partial q_2}{\partial t} + \frac{\partial f}{\partial q_3}\cdot\frac{\partial q_3}{\partial t} + \frac{\partial f}{\partial p_1}\cdot\frac{\partial p_1}{\partial t} + \frac{\partial f}{\partial p_2}\cdot\frac{\partial p_2}{\partial t}$$

$$+\frac{\partial t}{\partial p_3}\cdot\frac{\partial p_3}{\partial t}$$

$$= \frac{\partial f}{\partial t} + \frac{\mathrm{d}\boldsymbol{r}}{\mathrm{d}t}\cdot\frac{\partial f}{\partial \boldsymbol{r}} + \frac{\mathrm{d}\boldsymbol{p}}{\mathrm{d}t}\cdot\frac{\partial f}{\partial \boldsymbol{p}}$$

$$= \frac{\partial f}{\partial t} + \frac{\boldsymbol{p}}{m}\cdot\frac{\partial f}{\partial \boldsymbol{r}} + \boldsymbol{F}\cdot\frac{\partial f}{\partial \boldsymbol{p}} \tag{7.3}$$

which is the Eulerian, or convective, derivative of f. Remember that f here is *not* normalized: it represents the actual number distribution.

Now we turn to (2). For the moment, we denote the change of the number of particles in $\mathrm{d}\mu$ by $(\partial f/\partial t)_{\mathrm{c}}\cdot\mathrm{d}\mu$, the collision change, and we take this as a gain in the number of particles, since the sign will appear explicitly in a detailed calculation.

Then, the particle balance equation reads, over time Δt.

$$[f(\boldsymbol{r}, \boldsymbol{p}, t+\Delta t) - f(\boldsymbol{r}, \boldsymbol{p}, t)]\ \mathrm{d}\mu = (\partial f/\partial t)_{\mathrm{c}}\ \mathrm{d}\mu\ \Delta t$$

$$\frac{\partial f}{\partial t} + \frac{\boldsymbol{p}}{m}\cdot\frac{\partial f}{\partial \boldsymbol{r}} + \boldsymbol{F}\cdot\frac{\partial f}{\partial \boldsymbol{p}} = \left(\frac{\partial f}{\partial t}\right)_{\mathrm{c}} \qquad 7.4$$

since on dividing by $\mathrm{d}\mu\ \Delta t$ and taking the limit $\Delta t \to 0$ the l.h.s. is merely $\mathrm{d}f/\mathrm{d}t$.

Equation 7.4 is the Boltzmann transport equation. To use it, we must obtain some much more informative form for the r.h.s. Before discussing this, however, notice that, if there are no collisions, the r.h.s. must be zero and

$$\frac{\partial f}{\partial t} + \frac{\boldsymbol{p}}{m}\cdot\frac{\partial f}{\partial \boldsymbol{r}} + \boldsymbol{F}\cdot\frac{\partial f}{\partial \boldsymbol{p}} = 0 \qquad 7.5$$

This equation is widely used in plasma physics, where it is called the 'collisionless' Boltzmann or, often the 'Vlasov' equation. Since $(\partial f/\partial t)_{\mathrm{c}}$ might equal zero when the collisions are such that equal numbers of particles are scattered into and out of the test volume, or it might, at least, be a negligibly small difference between large numbers, the latter title seems more appropriate. We shall discuss the Vlasov equation in Chapter 8.

The form of the r.h.s. of equation 3.4 depends on the type of collision envisaged: between mass particles, between charged particles, between electrons, etc. We can consider, first, elastic collisions between mass particles in which momentum and energy are conserved:

$$\boldsymbol{p}_1 + \boldsymbol{p}_2 = \boldsymbol{p}_3 + \boldsymbol{p}_4 \qquad 7.6$$

$$p_1^2 + p_2^2 = p_3^2 + p_4^2 \qquad 7.7$$

The differential cross-section for scattering into an element of solid angle $\mathrm{d}\Omega$ is defined through

$$\mathrm{d}\sigma = \sigma(C, \theta)\ \mathrm{d}\Omega \qquad 7.8$$

where the cross-section σ is a function of the relative velocities of the two particles $|v_1 - v_2| = C$ and the angle of scattering θ, which itself is a function of v_1 and v_2.

The result, derived in Appendix 6, is

$$(\partial f/\partial t)_{\mathrm{c}} = \iint C\,\sigma(C,\theta)\ [f_3 f_4 - f_1 f_2]\ \mathrm{d}\boldsymbol{p}\ \mathrm{d}\Omega \qquad 7.9$$

Here f_1 is an abbreviation for $f(\boldsymbol{p}_1, \boldsymbol{r}, t)$ etc., and is the density of particles with momentum $\boldsymbol{p}$ in the volume element at $(\boldsymbol{r}, t)$. Then

$$\frac{\partial f}{\partial t} + \frac{\boldsymbol{p}}{m}\frac{\partial f}{\partial \boldsymbol{r}} + \boldsymbol{F}\cdot\frac{\partial f}{\partial \boldsymbol{p}} = \iint C\sigma\,[f_3 f_4 - f_1 f_2]\ \mathrm{d}\boldsymbol{p}\ \mathrm{d}\Omega \qquad 7.10$$

or we can rewrite this using the velocity, getting

$$\frac{\partial f}{\partial t} + \mathrm{v}\cdot\frac{\partial f}{\partial \boldsymbol{r}} + \frac{\boldsymbol{F}}{m}\cdot\frac{\partial f}{\partial \mathrm{v}} = \iint C\sigma\,[f_3 f_4 - f_1 f_2]\ \mathrm{dv}_1\ \mathrm{d}\Omega \qquad 7.11$$

The Boltzmann equation is therefore an integro-differential equation.

7.2 Boltzmann's H theorem

Boltzmann made an important observation about the behaviour of the function H defined by

$$H = \int f_2 \ln f_2 \, d\mathbf{v}_2 \qquad 7.12$$

which is the entropy, except for the constant $-k$. H is a function of t, but not of position. Then

$$\frac{\partial H}{\partial t} = \int (1 + \ln f_2) \frac{\partial f_2}{\partial t} \, d\mathbf{v}_2$$

$$= \iiint (1 + \ln f_2) \, [f_3 f_4 - f_1 f_2] \, C\sigma \, d\mathbf{v}_1 \, d\mathbf{v}_2 \, d\Omega \qquad 7.13$$

By permuting the f's, we can see that equation 7.13 is also

$$\frac{\partial H}{\partial t} = \tfrac{1}{4} \iiint (1 + \ln f_1 + 1 + \ln f_2 - 1 - \ln f_3 - 1 - \ln f_4) \, [f_3 f_4 - f_1 f_2] \, C\sigma \, . \, d\mathbf{v}_1 \, d\mathbf{v}_2 \, d\Omega$$

$$= \tfrac{1}{4} \iiint \ln\left(\frac{f_1 f_1}{f_3 f_4}\right) [f_3 f_4 - f_1 f_2] \, C\sigma \, d\mathbf{v}_1 \, d\mathbf{v}_2 \, d\Omega \qquad 7.14$$

If $f_1 f_2 > f_3 f_4$, the first term in the integral is positive, the second negative; while if $f_3 f_4 > f_1 f_2$, the first term is negative, the second positive. Therefore $\partial H/\partial t$ is always either zero or negative, and therefore H can never increase with time.* Thus, the *entropy* always increases or remains constant. It should be emphasized that this result applies only to an enclosed system of elastic mass particles and not necessarily to more general systems. Failure to appreciate this lack of generality was at the root of the gloomy late-Victorian forecasts of the thermal death of the universe, although Boltzmann himself strongly criticized such ideas.

Consider now a non-equilibrium state in which H is changing. H will decrease towards a limit in which $\partial H/\partial t = 0$ and from equation 7.14, this means that $f_3 f_4 = f_1 f_2$ or equally that $\ln f_1 + \ln f_2 = \ln f_3 + \ln f_4$. These relationships are independent of C. From the first form of equation 7.13 $\partial H/\partial t = 0$ also involves $\partial f_2/\partial t = 0$. Therefore, when the time required to reach equilibrium has elapsed, the gas is in a state which is both steady in time and uniform in space, according to the initial assumption on H. Furthermore, it can readily be shown that the equilibrium is such that every class of collison is exactly balanced by an inverse process, another example of detailed balancing.

While the H theorem is both interesting and historically important, as well as being of enormous practical importance in thermodynamics, it has also a great bearing on modern kinetic theory. We shall find, when we discuss plasma kinetics, that many different forms have been proposed for the collision integral on the r.h.s. of equation 7.11. These are developed for physical situations which are really outside

*Equation 7.14 is derived in a more formal way by Chapman and Cowling[4].

the boundaries of equilibrium statistical mechanics and are trying to describe complicated phenomena in a way which, it is hoped, will be tractable mathematically. However, if we can establish an *H* theorem for the proposed collision integral, we can conclude that it is well founded in the sense that it should correctly describe an equilibrium situation. Of course, we need to remember that, if the *H* theorem is invalid for the form derived, the latter may still be useful for non-equilibrium conditions.

We can derive more extremely useful information from equation 7.11 by the following procedure. We choose an arbitrary function of velocity $\phi(v_2)$, multiply equation 7.11 by $\phi(v_2)$, and integrate over v_2 to obtain the modified collision integral

$$\int \phi(\mathbf{v}_2) . C\sigma\, [f_3 f_4 - f_1 f_2]\ \mathrm{d}\mathbf{v}_1\, \mathrm{d}\mathbf{v}_2\, \mathrm{d}\Omega$$

Clearly it is allowable to interchange $\mathbf{v}_1$ and $\mathbf{v}_2$ without altering the value of the integral, so

$$\int \phi(\mathbf{v}_2) I \mathrm{d}\mathbf{v}_2 = \int \phi(\mathbf{v}_1) I \mathrm{d}\mathbf{v}_1$$

where I is shorthand for the Boltzmann integrand.

Next, we repeat the process with $\mathbf{v}_3$ and $\mathbf{v}_4$ and use the Jacobian of the transformation from $\mathbf{v}_1$, $\mathbf{v}_2$ co-ordinates to $\mathbf{v}_3$, $\mathbf{v}_4$ co-ordinates to show that

$$\int \phi(\mathbf{v}_3)(-I)\, \mathrm{d}\mathbf{v}_3 = \int \phi(\mathbf{v}_4)(-I) \mathrm{d}\mathbf{v}_4 = -\int \phi(\mathbf{v}_2) I\, \mathrm{d}\mathbf{v}_2$$

$$\text{or} \quad \int \phi(\mathbf{v}_2) I\, \mathrm{d}\mathbf{v}_2 = \tfrac{1}{4}\int\ [\phi_1 + \phi_2 - \phi_3 - \phi_4]\ I\, \mathrm{d}\mathbf{v}_2 \qquad 7.15$$

The importance of equation 7.15 resides in the fact that, when $\phi(\mathbf{v})$ represents a quantity which is conserved in a collision, $\phi_1 + \phi_2 = \phi_3 + \phi_4$ and equation 7.15 is zero. From the simple dynamics of binary collisions, there are five such quantities; the mass, the three components of momentum, and the energy. Then, we can write the following defining equations for the distribution function f:

$$\int Mf\, \mathrm{d}\mathbf{v} = NM \qquad 7.16$$

$$\int \mathbf{v} f\, \mathrm{d}\mathbf{v} = 0 \qquad 7.17$$

$$\tfrac{1}{2}\int Mv^2\ f \mathrm{d}\mathbf{v} = \text{total energy} \qquad 7.18$$

where the momentum equation has been written for the gas at rest (we could equate it to a constant for more generality). Remember now that f satisfies equation 5.11 with r.h.s. zero as well as these equations. Next, form a function $f \ln f + aM + bM\mathbf{v} + cMv^2$ and vary it with respect to f, (a, b and c being Lagrange multipliers). Then

$$\int \mathrm{d}\mathbf{v}\ [1 + \ln f + bM\mathbf{v} + cMv^2]\ \delta f = 0$$

which can be true only if the quantity in the square bracket is zero.

Therefore

$$\ln f = -(bM\,\mathrm{v} + cMv^2 + 1)$$

or $$f = K\exp(-cMv^2 - bM\mathbf{v}) \qquad 7.19$$

But we have the three equations 7.16 to 7.18 to determine the constants, and it is not difficult to show that the result is

$$f_{\mathrm{M}} = N\left(\frac{M}{2\pi kT}\right)^{\frac{3}{2}} \exp\left(-\frac{M\mathrm{v}^2}{2kT}\right) \qquad 7.20$$

or, if the average momentum is *not* zero but $M\langle u\rangle$,

$$f_{\mathrm{L}} = N\left(\frac{M}{2\pi kT}\right)^{\frac{3}{2}} \exp\left[\frac{-M(\mathrm{v}-u)^2}{2kT}\right] \qquad 7.21$$

Thus we have found that Maxwell's equation satisfies the Boltzmann equation identically.

Chapman and Cowling* give a stronger argument showing, furthermore, that this is a unique solution. This is extremely important, because it establishes the following. If we start with an isolated ensemble of gas particles which has been newly prepared and is not in equilibrium, then collisions will, in fact, operate so that the distribution function finally reaches the result given by equation 7.20 or 7.21. In this process H will decrease until equilibrium is reached, when it reaches its minimum value. This means that the Maxwell distribution is the fundamental reference distribution for all free particle flows which can be described in classical particle terms. Once the gas reaches equilibrium, the collisions continue to occur but now detailed balancing operates to ensure that, on average, the distribution function remains unchanged, although small short-time fluctuations are allowed. The reader will find a very interesting discussion giving the modern viewpoint of the significance of entropy, and therefore of H, in Levitch[5].

7.3 The relaxation-time approximation to the collision integral

The direct application of the collision integral, equation 7.9, is fraught with extreme difficulty, and it is advisable to gain familiarity with the Boltzmann transport equation in a simpler context which gives many practical results without much work. The central idea is that of perturbation theory: it is assumed that the instantaneous distribution function has been changed away from the equilibrium (Maxwellian) distribution by a small perturbation. The perturbation is, however, not in the nature of a fluctuation about the mean, but is caused by a defined outside influence on the situation. For example, one might consider what happens when a pulse of fairly high energy electrons is injected into a cool plasma. The distribution function of

*Reference 4: section 3.22 on summational invariance; section 4.1 for a more thorough derivation of the Maxwell distribution.

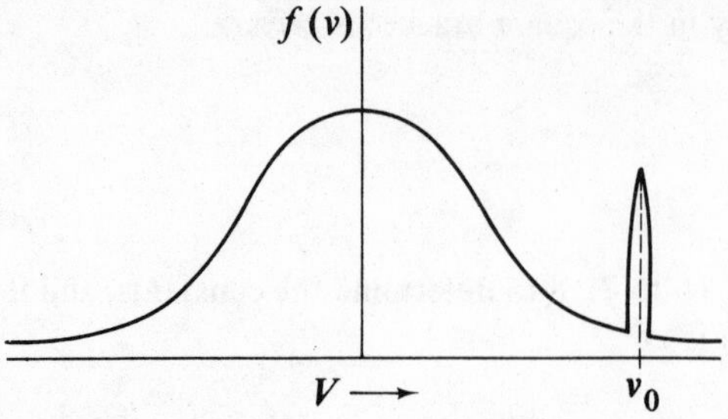

Fig. 7.1 A Maxwellian plasma penetrated by a practically monoenergetic particulate beam.

electrons (plasma electrons + injected electrons) is then as shown in Fig. 7.1, and one might wish to study the manner in which a new equilibrium is established. A more mundane example is the calculation of the electrical conductivity of a semiconductor (in which the conduction electrons obey Maxwell–Boltzmann statistics) so that the equilibrium distribution is distorted by a field acting along the semiconductor. It is then *assumed* that collisions will act so as to restore the equilibrium distribution function f_0. It is not necessary to be very precise about the nature of the collisions, but it is assumed that they will restore the equilibrium in an exponential fashion with a time constant τ_1 which is related to the time *between* collisions.

If several types of collision take place in the system, an effective relaxation time

$$\frac{1}{\tau_{\text{eff.}}} = \frac{1}{\tau_{e1}} + \frac{1}{\tau_{e2}} + \frac{1}{\tau_{e3}} + \text{etc.} \qquad 7.22$$

can be defined. Often, $\tau_{e1} \ll \tau_{e2} \ll \tau_{e3}$, so that one kind of collision is dominant. The reader will appreciate that the concept of relaxation time is indefinite and, in practice, has to be developed from a more detailed consideration of the collision processes for particular cases. In spite of this, the concept is extremely valuable.

In view of the assumptions made above, we put

$$(\partial f/\partial t)_c = (f_0 - f)/\tau_e \qquad 7.23$$

where f is the perturbed distribution and f_0 is the equilibrium distribution. For fixed r and p we can use equation 7.3 to get

$$df/dt = (f_0 - f)/\tau_e$$

$$\text{or} \quad f(t) = [f(O) - f_0(O)] \exp(-t/\tau_e) + f_0(t) \qquad 7.24$$

confirming the exponential character of the relaxation.

Equation 7.24 has introduced a time variation into the distribution function in a natural way. The method is important and we now discuss examples of its use.

Example 1. Electrical conductivity of semiconductors and gases

A semiconducting rod, long in the z direction, is considered, so that the Maxwell–Boltzmann distribution is relevant. A uniform electric field E_z is present when a steady-state current flow has been established and this flow is uniform along the rod. Then $\partial f/\partial t = 0$ and $\partial f/\partial z = 0$, so that equations 7.3 and 7.23 reduce to

$$eE_z \frac{\partial f}{\partial p_z} = \frac{f_0 - f}{\tau_e} \qquad 7.25$$

If E_z is small, we can put $f = f_0 + f_1, f_1 \ll f_0$

therefore $eE_z \dfrac{\partial f_0}{\partial p_z} = -\dfrac{f_1}{p_z}$

as a first approximation to equation 7.25.

$$\text{Then} \quad f_1 = f - f_0 = -eE_z \tau \frac{\partial f_0}{\partial p_z} \qquad 7.26$$

$$\text{But} \quad \frac{\partial f_0}{\partial p_z} = -\frac{p_z f_0}{mkT}$$

$$\text{therefore} \quad f_1 = f_0 \left(1 + e\frac{E_z \tau_e p_z}{mkT}\right) \qquad 7.27$$

This result shows that more electrons travel in the positive direction of eE_z than in the E_z negative direction. It is therefore physically reasonable.

To complete the calculation, we note that the current element is $\Delta I = ep_z/m$ so that a density of N_e electrons will give rise to a current density

$$J_z = eN_e \int_0^{\infty} \iint_{-\infty}^{+\infty} f_0 \left\{\frac{p_z}{m} + \frac{2eE_z\tau_e}{mkT}\left(\frac{p_z}{2m}\right)^2\right\} \mathrm{d}v_p$$

where v_p means v_x, v_y, v_z according to which integral we are evaluating. But we can avoid actually doing the integrals in detail if we remember that

$$\iiint \frac{p_z}{m} f_0 \, \mathrm{d}v = \frac{1}{m}\langle p_z \rangle = 0$$

$$\iiint \frac{p_z^2}{2m} f_0 \, \mathrm{d}v \left\langle \frac{p_z^2}{2m} \right\rangle = \frac{kT}{2}$$

$$\text{Then} \quad J_z = N_e \cdot \frac{e^2 E_z \tau_e}{m}$$

It would be better to use the effective mass m^* in lieu of m, so that the final result is

$$J_z = N_e \cdot \frac{e^2 E_z \tau_e}{m^*}$$

$$\text{or} \quad \sigma = N_e \cdot \frac{e^2 \tau_e}{m^*} \qquad 7.28$$

Equation 7.28 includes temperature variation, mainly because N_e is a rapidly increasing function of temperature while τ_e varies slowly with temperature. The agreement with experiment is good enough to strengthen our confidence in the relaxation method.

Equation 7.28 is directly applicable to the electronic and ionic conductivities of a weakly ionized gas (one in which $N_i, N_e \ll N_0$ the molecular density) provided τ_e and m are properly redefined. It immediately shows that the ionic conductivity is smaller than the electronic conductivity in the ratio m/M_i, a large number.

Example 2 The alternating conductivity

The physical system is the same as that of the first example, with the exception that an alternating field $E_z \exp(j\omega t)$ is applied. Since the perturbation must now alternate at the same frequency as the field, we must put $f = f_0 + f_1 \exp(j\omega t)$, giving $\partial f/\partial t = j\omega f_1 \exp(j\omega t)$.

Then $$eE_z \frac{\partial f_0}{\partial p_z} = -f_1\left(\frac{1}{\tau_e} + j\omega\right) \qquad 7.29$$

In treatments of plasma physics, it is usual to denote $1/\tau_e$ by ν_c – the 'collision frequency' – and this saves writing, so we use the convention here.

The reader can easily fill in the details for himself, as they are identical with the earlier treatment. The final result is

$$\sigma = \frac{N_e \, . \, e^2}{m^*}\left(\frac{1}{\nu_c + j\omega}\right)$$

$$= \sigma_0 \frac{(1 - j\omega\tau_e)}{[1 + (\omega\tau_e)^2]} \qquad 7.30$$

where σ_0 is the zero frequency conductivity

or $$\rho = \rho_0(1 + j\omega\tau_e) \qquad 7.31$$

Thus, the a.c. impedance is that of an inductive resistor and departs noticeably from the d.c. value when $\omega\tau_e$ is 0(1). In principle, therefore, equation 7.31 gives an experimental method for the measurement of τ_e, provided that this time is not so short that it is not practicable to reach large enough values of ω.

It is not difficult to extend the treatment to the three-dimensional case, obtaining the components of the conductivity tensor, and to include a magnetic field. These matters are developed in most books on plasma physics and the reader will find treatments in references 6 and 7. We return to the question in Chapter 8.

These examples are particularly relevant to the themes of this book, but many other properties, for example the viscosity and the thermal conductivities of a gas, can be calculated by the same technique. The technique can also be applied to systems in which the equilibrium is *not* Maxwellian; for example, the conductivity of metals can be computed for f_0 of Fermi–Dirac form, and we shall carry out the details as another exercise.

To conclude this discussion we might ask the question, what limits the validity of the development? In the derivation of example 1, we assumed $f_1 \ll f_0$. But equation 7.27 shows that this means that $eE_z\tau_e p_z/mkT \ll 1$, with p_z of order $(mkT)^{\frac{1}{2}}$. Then, the condition is that $eE_z\tau_e \ll p_z$ or that the energy gained by a particle in one relaxation time is much less than the mean thermal momentum. It is well known that it is difficult to establish strong fields inside real, good conductors, so this restriction is not encountered in usual experimental conditions. In other examples, similar conditions can be isolated which may or may not be restrictive in these instances. It is worth pointing out that a better approximation, instead of equation 7.27, could be obtained by iteration; that is, one could put equation 7.27 for f in equation 7.25 to obtain a more accurate approximation.

Example 3 Electrical conductivity of metals

The same basic technique can be used to obtain a reasonably good approximation for the conductivity of metals, if we use the distribution function of Fermi–Dirac instead of the Maxwellian form. Equation 7.25 still applies, and we reorganize it as

$$f_1 = -\frac{e\tau_e\, \partial f_0/\partial p_z}{(1 + \mathrm{j}\omega\tau_e)} \,.\, E(z)$$

Both f_1 and $E(z)$, of course, vary as $\exp(\mathrm{j}\omega t)$. To save writing, we put $\tau_e{}' = \tau_e$ $(1 + \mathrm{j}\omega\tau_e)$ and the conductivity is obtained by multiplication of f_1 by $-2ev_z$ and integrating over all v_z. Therefore

$$\sigma = -\frac{2e^2}{m^*}\int \tau_e{}'\, v_z\left(\frac{\partial f_0}{\partial v_z}\right)\mathrm{d}v_z \qquad 7.32$$

In general, the evaluation of equation 7.32 is complicated, but it is very easy if we assume that the Fermi surface in the metal is a sphere, i.e. that the distribution is isotropic. Then, from Fig. 4.3, we remember that $\partial f_0/\partial v_z$ has the character of a delta function, and the relaxation time τ_e, which in general is a function of velocity, can be assumed to take on the constant value relevant to the Fermi surface, which we call τ_F. Then, we can approximate equation 7.32 by

$$\sigma = -\frac{2e^2\tau_F'}{m^*}\int v_z\frac{\partial f_0}{\partial v_z}\,\partial v_z \qquad 7.33$$

Integrating by parts, the integral becomes

$$\Big[f_0\, v_z\Big]_{-\infty}^{+\infty} - \int_{-\infty}^{+\infty} f_0\,\mathrm{d}v_z = -\int_{-\infty}^{+\infty} f_0\,\mathrm{d}v_z \qquad 7.34$$

where the last result comes from the fact that $f_0 \to 0$ at the limits. The last integral is the z component of $\int f_0\,\mathrm{d}v$, but this integral is simply N_e, the density of electrons per unit volume. Finally, then,

$$\sigma = -\frac{2e\,N_e}{m^*}\,\tau_F{}' \qquad 7.35$$

The significance of this result is better brought out by writing $\tau_F = \lambda_F/u_F$ where λ_F is the mean free path and u_F is the velocity on the Fermi surface. Equation 4.29 shows that N_e is a function of u_F as well as depending on temperature. Moreover, it can be shown, by measurements of Hall coefficient and conductivity over a range of temperatures, that λ_F is a decreasing function of temperature which is much greater in magnitude than the interatomic spacing. Reference 8 can be consulted for full details.

More accurately, we should write the collision term as

$$\frac{\lambda_F}{u_F}\left(\frac{1}{1+j\omega\,\lambda_F/u_F}\right)$$

but the imaginary part is very small for all ordinary frequencies. If we take λ_F as $O(10^{-6}\,\text{m})$ and u_F as $O(10^{6}\,\text{m/s})$, τ_F is $O(10^{-12})$ and the imaginary term is unity for $\omega = 10^{12}$. Thus, for all ordinary purposes, the conductivity is *real*, and observed complex terms are due to the ordinary, macroscopic skin effect.

7.4 Perturbation solution of Boltzmann's equation

We saw in Chapter 2 that the calculation of many physical parameters of gases depended on a calculation of the flux of particles, or momentum, or energy across an imaginary surface drawn in the gas. In this section we seek to make more accurate calculations of the same type. The general problem is called the 'transport' problem and it has been brought to a very high degree of development in several fields, one of the most noteworthy being the exhaustive treatment of non-ideal gases by Chapman and Cowling[4]. The reader will find more modern treatment, from different points of view, in the books by Kogan[8] and by Cercignani[10].

Here, we attempt to expose only the more important ideas of this approach, without going into the fine detail of modern gas dynamics.

The first step is to set up the equation named after Enskog, who initiated many developments in the modern kinetic theory of gases, and after Chapman who developed the method.

The method is a perturbation one, which like all such methods is quite easy to describe in principle but demands much detailed calculation for practical use. It is the basic method used by Chapman and Cowling[4] in their exhaustive account of transport theory in moderately dense gases. For this reason, and because it is not widely useful in charged-particle theory, we shall only point out the main features here.

First, we return for a while to the different forms of Maxwellian distribution defined in equations 7.20 and 7.21. Equation 7.20 defines the final, equilibrium distribution which must be reached in a gas in which the temperature and particle number density are uniform throughout the whole volume considered and, from the derivation, there is no average translation of the gas. In equation 7.21, on the contrary, closer examination shows that the so-called local or drifting Maxwellian there defined implies that T and N are functions of position and, therefore, of time, so as to satisfy equations of continuity. For example, this could clearly be the case for gas flow along a long pipe of variable section. In principle, one can base a perturbation theory on either equation 7.21 or on equation 7.21, but the results will not be the same and it is important to understand the physics of the difference. If the gas as a whole is nearly in a state of complete equilibrium, it is clearly advisable to base the perturbation on equation 7.20 since the disturbance ought to relax to equation 7.20. In this case, we would put

$$f(r, \mathrm{v}, t) = f_M\,[1 + \zeta(r, \mathrm{v}, t)] \qquad 7.36$$

where ζ is a small perturbation term, $\ll 1$.

Secondly, we might use

$$f(r, \mathrm{v}, t) = f_L\,[1 + \epsilon(r, \mathrm{v}, t)] \qquad 7.37$$

where again $\epsilon \ll 1$. This is the *Chapman–Enskog* form. Analysis shows that it describes the behaviour of the distribution for times which are greater than the characteristic time τ defined by the ratio of the mean free path of a molecule to the mean molecular velocity. In symbols, τ is $O(\lambda/\langle u \rangle)$. Here, the idea is that, if the distribution function is changing with position, it is only sensible to describe its behaviour at time intervals large enough to allow several collisions to have taken place.

The first step in the Chapman–Enskog method is to put equation 7.37 into the Boltzmann transport equation, writing, in the collision integral,

$$f_n = f_L\,[1 + \epsilon_n]$$

The result is

$$\frac{\partial f_n}{\partial t} + v_k \frac{\partial f_L}{\partial x_k} + F_k \frac{\partial f_L}{\partial v_k} = f_L \int C\sigma f_{L1}(\epsilon_3 + \epsilon_4 - \epsilon_1 - \epsilon_2)\, dv_1\, d\Omega \qquad 7.38$$

(For simplicity, this is written in tensor notation, but we shall not need to use tensor analysis now). Equation 7.38 still looks extremely difficult, but compared with the Boltzmann transport equation it has the great advantage of linearity. Mathematically speaking, it is a linear, inhomogenous integro-differential equation, and a formidable array of apparatus for the solution is available.

One of these techniques is worth mentioning before we embark on further description of the methods used by Chapman and Enskog, because we shall refer to a similar technique again and because it uses the variational methods which, in the last decade, have been developed to a very high degree of sophistication in other fields of engineering and physics and which lend themselves to machine computation.

If equation 7.38 is written as an operator equation, it can be expressed as

$$\mathrm{L}\epsilon = A \qquad 7.39$$

where L is a linear integral operator and A is known. Then, if an arbitrary function ψ is introduced, which solves the equation

$$\int \psi\, \mathrm{L} \psi\, d\mathrm{v} = \int \psi A \psi\, d\mathrm{v} \qquad 7.40$$

it is possible to show that

$$\int \epsilon \, \mathrm{L} \epsilon \, \mathrm{d}v \geqslant \int \psi \, \mathrm{L} \psi \, \mathrm{d}v \qquad 7.41$$

In other words, the solution is that function satisfying equation 7.40 which maximizes the integral $\int \epsilon \mathrm{L} \epsilon \mathrm{d}v$. Then, it only remains to express ϵ as a series $\Sigma a_i g_i$, adjusting the coefficients a_i until the maximum is achieved, the g_i being chosen functions. The whole problem then is very similar to the use of variational techniques to determine, for instance, the fields in complex electromagnetic structures.

The Chapman–Enskog technique is quite different. It extends the perturbations to higher orders, using the properties of the first-order solution to improve the solution of the second stage and so on. To set out the method involves a good deal of special notation, to prevent the equations from becoming excessively cumbersome. We use a notation loosely based on Chapman's. The original equation is written in the form

$$\xi(F) \doteq \frac{\mathrm{D}F}{\mathrm{D}t} + \hat{J}F = 0 \qquad 7.42$$

Here $\xi(F)$ is an operation on the (unknown) function F, the first term on the l.h.s. is the convective derivative and $\hat{J}$ is the shorthand for the collision integral as originally defined, in equation 7.11 for example. Then, assume that

$$F = F_0 + F_1 + F_2 + \ldots + \text{etc.} \qquad 7.43$$

Furthermore, suppose that when this series is subjected to the operation ξ the result is such that rth term involves *only* the first r terms in equation 7.43. Then

$$\xi(F) = \xi_0(F_0) + \xi_1(F_0, F_1) + \xi_2(F_0, F_1, F_2) \text{ etc.} \qquad 7.44$$

One way of satisfying $\xi(F) = 0$, equation 7.42, is to require that

$$\xi_0(F_0) = 0 \qquad 7.45$$

$$\xi_1(F_0, F_1) = 0 \qquad 7.46$$

and so on. Then, if F_0 can be found from equation 7.45, F_1 can be found from equation 7.46 and so on along the whole hierarchy of equations. This division is a very special one and, *ab initio*, might even be impossible, but Chapman and other authors have carefully investigated the problem with the result that the division can be justified.

To proceed, substitute equation 7.43 into the middle term of equation 7.42. The result is

$$\xi(F) = \frac{\mathrm{D}}{\mathrm{D}t}\left(\sum_r F_r\right) + \hat{J}\left\{\left(\sum_r F_r\right)\left(\sum_s F_{1s}\right)\right\} \qquad 7.47$$

where the notation shows that the collision operator depends on *two* distribution functions and their modification by collisions. The next stage is singled out by Chapman as due to Enskog, and is responsible for the power of the method. His expansion of $\hat{J}$ in equation 7.47 groups the terms in the manner used in grouping terms when expressing the product of two series as another series, viz.

$$\sum(x_0 y_r + x_1 y_{r-1} + \ldots + x_r y_0)$$

Also, D/Dt is divided into time and space parts, the time variation being considered later on. Writing D for the space part, Enskog puts

$$D_0 = 0 \qquad 7.48$$

and allows D_r to depend only on $F_0, \ldots F_{r-1}$. Then, if we write the rth term of equation 7.47 as

$$\xi_r(F) = \hat{J}_r + D_r$$

we find $\xi_0(F) = \hat{J}_0 = 0$

from equations 7.45 and 7.48

and $\xi_r(F) = \hat{J}_r + D_r = 0 \qquad 7.49$

from the rth term of equation 7.46.

Now, we have already shown in the discussion leading to equation 7.21 that f_L as there defined makes $\hat{J}_0 = 0$. Thus, the local Maxwellian is certainly an allowable solution for F_0. We can therefore write out the whole solution so far as $F_0 = f_L$

and $$\int F_0 \begin{pmatrix} 1 \\ \mathbf{v} \\ v^2 \end{pmatrix} d\mathbf{v} = \begin{pmatrix} N \\ N\langle v \rangle \\ 3NkT/m \end{pmatrix} \qquad 7.50$$

The information available does not allow the pressure tensor or the heat flow to be calculated, so we therefore see that to get beyond elementary results at least one further stage of perturbation is required.* We must then find a solution for F_1 from

$$\xi(F_1) = D_1 + \hat{J}_1$$

which is equation 7.49 for $r = 1$. Also, D_1 depends only on D_0 as stated. A sketch of the procedure is as follows (the full detail is given in Chapman and Cowling[3], section 7.3). For D_1, the l.h.s. of the Boltzmann equation written in terms of the random velocity $\{\mathbf{v} - \langle v \rangle\} = c$ and the translational velocity c_0 is used, i.e.

$$D_1 = \frac{D_0 F_0}{Dt} + c\frac{\partial F_0}{\partial r} + \left(F - \frac{D_0 c_0}{Dt}\right) \cdot \frac{\partial F_0}{\partial c} - \frac{\partial F_0}{\partial c} \cdot c \cdot \frac{\partial c_0}{\partial r} \qquad 7.51$$

Vector force F should not be confused with the functions F.

Equation 7.51 can be modified by eliminating force in terms of pressure, and we can operate on ln F_0 instead of on F_0.

$$D_1 = F_0 \left\{ \frac{D_0 \ln F_0}{Dt} + c\frac{\partial \ln F_0}{\partial r} + \frac{1}{\rho} \cdot \frac{dp}{dr} \cdot \frac{\partial \ln F_0}{\partial c} - \frac{\partial \ln F_0}{\partial c} \cdot c \cdot \frac{dc_0}{dr} \right\} \qquad 7.52$$

where ρ = density, p = pressure.

*The rest of this section may be omitted on a first reading.

But we know that

$$\ln F_0 = \text{const.} + \ln\left(\frac{N}{T^{\frac{3}{2}}}\right) - \frac{mc^2}{2kT} \tag{7.53}$$

by elimination of the mean thermal velocity from the second result of equation 7.50. Therefore

$$\frac{\partial \ln F_0}{\partial c} = -\frac{mc}{kT} \tag{7.54}$$

Further,

$$\frac{\mathrm{D}_0 \ln F_0}{\mathrm{D}t} = \frac{mc^2}{2kT^2} \cdot \frac{\mathrm{D}_0 T}{\mathrm{D}t}$$

which, after several manipulations, can be shown to equal

$$-\frac{mc^2}{3kT}\frac{\partial}{\partial \mathrm{v}} \cdot c_0$$

since it can be shown that

$$\frac{\mathrm{D}_0}{\mathrm{D}t}\left(\frac{n}{T^{\frac{3}{2}}}\right) = 0$$

Futhermore, the second and third terms in the curly bracket of equation 7.52 simplify to $\left(\frac{mc^2}{2kT} - \frac{5}{2}\right) c \frac{\partial \ln T}{\partial r}$ and, summing the other two, we find that the result is $\frac{m}{kT}\overline{\overline{cc}}\colon\frac{\partial c_0}{\partial r}$

where, if $c = u, v, w$ and $|c| = c$,

$$\overline{\overline{cc}} = \begin{bmatrix} u^2 - \frac{c^2}{3} & uv & uw \\ vu & v^2 - \frac{c^2}{3} & vw \\ wu & wv & w^2 - \frac{c^2}{3} \end{bmatrix}$$

$$\mathrm{D}_1 = F_0\left\{\left(\frac{mc^2}{2kT} - \frac{5}{2}\right) c \frac{\partial \ln T}{\partial r} + \frac{m}{kT} \cdot \overline{\overline{cc}}\colon\frac{\partial}{\partial r}(c_0)\right\} \tag{7.55}$$

$$\text{or} \quad F_0\left\{\left(\mathscr{C}^2 - \frac{5}{2}\right) c \frac{\partial \ln T}{\partial r} + 2\overline{\overline{\mathscr{C}\mathscr{C}}}\colon\frac{\partial}{\partial r}(c_0)\right\} \tag{7.56}$$

where $\mathscr{C}^2 = (m/2kT)\, c^2$ and is dimensionless.

In terms of $\mathscr{C}$, $F_0 = N(m/2\pi kT)^{\frac{3}{2}} \exp(-\mathscr{C}^2)$.

We next see that the integral part of $\hat{J}$ is given by

$$\hat{J}_1 = \hat{J}(F_0, F_{11}) + \hat{J}(F_1, F_{10})$$

then, since $\mathscr{C}$ is conserved in collision so that the development leading to equation 7.15 holds, we can put $F_1 = F_0\ \phi_1$ and find

$$\mathscr{J}_1 = N^2 \phi_1 \tag{7.57}$$

the $\frac{1}{4}$ in equation 7.15 having been taken into the definition of ϕ.

We now have $F = F_0(1 + \phi_1)$, and therefore

$$\int F_0 \cdot \phi_1 \begin{pmatrix} 1 \\ \mathrm{v} \\ \mathrm{v}^2 \end{pmatrix}' \mathrm{dv} = 0 \tag{7.58}$$

Equation 7.58 allows ϕ_1 to be determined in terms of two auxiliary quantities which are functions of N, T and $\mathscr{C}$. Further, and very elaborate, analysis determines the value of these auxiliaries, which are A and $\bar{\bar{B}}$ in

$$F = F_0 [1 + (2RT)^{\frac{1}{2}}\ A \,.\, \nabla \ln T + \bar{\bar{B}} \,.\, \bar{\bar{\nabla}} v] \tag{7.59}$$

The reader will now be able to appreciate the labour involved in reaching the third approximation, which Chapman and Cowling have done. The results obtained are valuable in advanced gas dynamics and in fields such as the use of the diffusion process for the separation of isotopes.

The most remarkable feature of the Chapman–Enskog development is the skilful way in which the conditions, which are vital if there is to be agreement with elementary gas kinetics, on the density, temperature, internal energy and, from the second approximation, on the viscosity and thermal conductivity, have been preserved and, in the third approximation, comparatively simple correction terms derived. The reader will probably be well aware that, while normal perturbation methods are useful for deriving numerical corrections, they do not normally provide information in an analytical form. We shall shortly see that this requirement of preserving the physical parameters is a basic consideration in the moment method of solving the transport equations.

Before taking up the moment problem we shall describe another perturbation method based on the equilibrium distribution $f_{(M)}$. Consider small perturbations away from the Maxwellian equilibrium, expressed by

$$f \approx f_{(M)}\ [1 + \epsilon] \tag{7.60}$$

Then

$$\frac{\partial \epsilon}{\partial t} + v_{\mathrm{k}} \frac{\partial \epsilon}{\partial q_{\mathrm{k}}} + \frac{F_{\mathrm{k}}}{m} \cdot \frac{\partial \epsilon}{\partial v_{\mathrm{k}}} = \int \mathscr{C}\sigma f_{(M)} (\epsilon_3 + \epsilon_4 - \epsilon_1 - \epsilon_2)\ \mathrm{dv} \mathrm{d}\Omega \tag{7.61}$$

because $f_{(M)}$ solves the unperturbed Boltzmann equation. Now, we specialize still more and consider the case when $F_{\mathrm{k}} = 0$ and therefore $v_{\mathrm{k}} = 0$. Then

$$\frac{\partial \epsilon}{\partial t} = \int \mathscr{C}\sigma f_{(M)} (\epsilon_3 + \epsilon_4 - \epsilon_1 - \epsilon_2)\ \mathrm{dv}_1\, \mathrm{d}\Omega$$

If we try

$$\epsilon(\mathrm{v}, t) = \sum_n A_n \psi_n \exp(-t/\tau_n(\mathrm{v})) \qquad 7.62$$

a linear integral equation results, which can be solved only if $\sigma(v)$ is known. Since this is not generally the case, it is not possible to get a complete solution, corresponding with several different relaxation times, each valid for a different velocity class.

What we can do, however, is to take one term only:

$$f = f_{(\mathrm{M})}\ [1 + \phi \exp(-t/\tau)]$$

Then $\dfrac{\partial \epsilon}{\partial t} = -\dfrac{\phi}{\tau} \exp\left(-\dfrac{t}{\tau}\right) = -\dfrac{\epsilon}{\tau}$

But $\epsilon \approx \dfrac{f - f_{(\mathrm{M})}}{f_{(\mathrm{M})}}$

therefore

$$\frac{\partial f}{\partial t} + f_{(\mathrm{M})} \frac{\partial \epsilon}{\partial t} = -\frac{(f - f_{(\mathrm{M})})}{\tau} = \frac{f_{(\mathrm{M})} - f}{\tau} \qquad 7.63$$

which is the previously assumed relaxation expression. The essential step in the derivation is realizing that the integral is independent of t so that expansion 7.62 is proper. Another piece of information emerges. We have also the relationship $\partial\epsilon/\partial t = I$, the r.h.s. of equation 7.61. Therefore we can write

$$\frac{1}{\tau} = \frac{f_{(\mathrm{M})} I}{f_{(\mathrm{M})} - f} \qquad 7.64$$

so that τ is defined in terms of the collision integral. Since this integral can sometimes be calculated fairly exactly, this is a useful technique for the determination of τ. By inspection of I, it is seen that I is of order $N \langle C\sigma \rangle$; therefore, the present section serves to clarify the physical meaning of the relaxation time, which was ill-defined when first introduced.

Before leaving equation 7.61, we note that it is linear, just as was equation 7.38, but it is also of a simpler type. It can be written, in operator form, as $\partial\phi/\partial x_i = \hat{\mathrm{L}}\phi$, Then, if the eigenvalues and eigenfunctions of $\hat{\mathrm{L}}$ are v_n, ϕ_n, we can expand $\hat{\mathrm{L}}\phi_n$ as

$$\hat{\mathrm{L}}\phi_n = v_n \phi_n \qquad 7.65$$

The result for ϕ is then

$$\phi = \sum_i a_i \phi_i \qquad 7.66$$

Now, it is important to remember that the five (scalar) collisional invariants reduce the r.h.s. of equation 7.61 to zero and are independent of the coordinates. In other words, they are eigenvalues of $\hat{\mathrm{L}}$. The a_1 are functions of t in the general case, so we can be a little more explicit, writing

$$\phi(\mathrm{v}, t) = \sum \beta_i \exp(-v_n t)\,\phi_n \qquad 7.67$$

Since the function may be put into orthonormal form, the inversion formula for the β_i becomes

$$\beta_i = \int \phi_0(\mathrm{v})\phi_n(\mathrm{v})\,\mathrm{d}v \qquad 7.68$$

where the suffix 0 denotes the time zero and a volume integration is implied. Chapman and Cowling[4] gives more details.

7.5 Moment methods and Boltzmann's equation

Moment methods for the solution of the Boltzmann equations stem from quite elementary ideas of fluid mechanics, although they have been brought to their present level of sophistication as a consequence of modern interest in kinetic theory.

To introduce the subject, let us write down the Boltzmann transport equation in the original form:

$$\frac{\partial f}{\partial t} + \mathrm{v}\frac{\partial f}{\partial r} + \frac{F}{m}\cdot\frac{\partial f}{\partial \mathrm{v}} = \left(\frac{\partial f}{\partial t}\right)_c$$

Then, as we have done before, we multiply by a function $\phi(\mathrm{v})$ and integrate in velocity. For the moment we ignore the r.h.s. but allow F to include both forces which do not depend on velocity, such as gravity or electric fields, and forces which do, including magnetic forces.

Exercise
Show that, for an electromagnetic field,

$$\frac{\partial}{\partial t}(N\langle\phi\rangle) + \frac{\partial}{\partial r}(N\langle\phi\mathrm{v}\rangle) - \frac{N_e}{m}[E + \nabla \times B]\left\langle\frac{\mathrm{d}\phi}{\mathrm{dv}}\right\rangle = 0 \qquad 7.69$$

We have become accustomed to thinking of equations of this sort as specifying the average when the distribution function is known. Now we are going to invert the problem, assuming that the average is known, from measurements or theory, and we wish to find f from this knowledge. We notice immediately the major difficulty in this programme. Suppose we know N, then we take $\phi = 1$ and look at

$$\int \phi \frac{\partial f}{\partial t}\,\mathrm{dv} = \frac{\partial}{\partial t}\int \phi f\,\mathrm{dv}$$

which can be solved only if we know $f(\mathrm{v})$. We therefore look at the integrals taken over $\phi = \mathrm{v}$, finding that we must know $\langle\mathrm{vv}\rangle$ and so on. Formally, we express this by saying that each moment equation involves the velocity moment next in ascending order. The method of overcoming this would appear to be to try to find a way of terminating the sequence at a high-orde- moment, which will still give accurate results for low orders.*

*Appendix 7 should be revised at this point.

The particular method due to H. Grad[1] is nowadays commonly used. Grad's technique is a clever blend of older-fashioned moment theory with a technique of polynomial expansion which was freely used by Chapman and Cowling. To improve the ease of writing, Grad defines the distribution function so that

$$\int \begin{pmatrix} g \\ g\mathrm{v} \\ g\mathrm{v}^2 \end{pmatrix} \mathrm{dv} = \begin{pmatrix} 1 \\ 0 \\ 3 \end{pmatrix} \tag{7.70}$$

where v is a non-dimensional velocity defined through

$$C = \zeta - u(x, t) \equiv \sqrt{(RT)}\,.\,\mathrm{v}$$

and g through

$$M = Nm = \int F(\zeta)\,\mathrm{d}\zeta\,\mathrm{d}x = \frac{\rho}{(RT)^{\frac{3}{2}}} \int g(\xi)\,\mathrm{d}(\xi)\,\mathrm{d}x$$

The Maxwellian is $F_M = \dfrac{\rho}{(RT)^{\frac{3}{2}}}\,\omega(v)$ in this notation.

The actual distribution function is expanded in a series of Hermite polynomials. The properties of these polynomials are discussed in detail in all textbooks on quantum mechanics, because they are the optimum orthogonal functions for the quantum-mechanical solution of the harmonic oscillator. Reference 11 contains a full account. Here, their particular merit is that they are excellently suited for the expansion of Maxwellians (or Gaussians) because the nth order Hermite polynomial is defined* as

$$H_n(x) = \frac{(-1)^n}{\exp(-x^2)} \frac{\mathrm{d}^n[\exp(-x^2)]}{\mathrm{d}x^n} \tag{7.71}$$

when x is a real number. In the form required by the analysis of the Boltzmann equation, H_n is defined as a tensor of order n, whose coefficients are polynomials of order n; thus

$$H^n_{ijk} = \frac{(-1)^n}{f_M} \cdot \frac{\mathrm{d}^n(f_M)}{\mathrm{d}v_i\,\mathrm{d}v_k\,\mathrm{d}v_l \ldots} \tag{7.72}$$

where f_M is the Maxwellian. Then we can expand F by putting

$$F = F_M \sum \frac{1}{n!} a_1^{(n)} H_1^{(n)}(\mathrm{v}) \tag{7.73}$$

which is equivalent to

$$g = \omega[a^0 H^0 + a^1_i H^1_i + \tfrac{1}{2} a^2_{ij} H^2_{ij} + \text{etc.}] \tag{7.74}$$

*Unfortunately, the definitions are not completely standardized, so care is necessary.

As usual, we can multiply this by H^0, H_i^1, etc. and integrate, using the orthogonality to find

$$a_i^{(n)} = \int g H_i^{(n)} \, dv = \frac{1}{\rho} \int F(\zeta) H_i^n \, d\zeta \qquad 7.75$$

Thus, the $a_i^{(n)}$ are the moments of $F(\zeta)$ and we have immediately

$$a^0 = 1$$
$$a_i^1 = 0$$
$$a_{ij}^2 = (P_{ij} - p\delta_{ij})/p$$
$$a_{ijk}^3 = 2Q_i/p\sqrt{(RT)}$$

These can then be put into the series for g (or F) and we can use this as a two- or three-term approximation for F in terms of F_0.

However, the technique is not limited to the collisionless case, which we have discussed so far. We shall not include the details, which are cumbersome but not difficult. Grad takes the collision in the same form as that of equation 7.15, replacing the ϕ's by Hermite polynomials in v. This is used to correct the a's defined above, which is easy to do since the order of each correction tensor determines which term it modifies.

The first approximation to F (F_M being the zeroth) is

$$F = F_M \left(1 + \frac{a_{ij}^2}{2} H_{ij}\right)$$

This, together with the equations of continuity and momentum, leads to a closed set. Perhaps the most common result found in the literature is the so-called 'thirteen-moment' expansion, in which the first thirteen moments having physical significance are retained. The result for F is

$$F = F_M \left[\left(1 + \frac{p_{ij}}{2N} \frac{v_i v_j}{(kT)^2}\right) - \frac{q_k v_k}{mN} \left(\frac{m}{kT}\right)^2 \left(1 - \frac{mv^2}{5kT}\right)\right] \qquad 7.76$$

The book by Kogan[9] discusses in detail the use of the method in gas dynamics.

7.6 Irreversibility and the Boltzmann equation

It would be unrealistic not to include a brief discussion of irreversibility, which is a consequence of the Boltzmann equation but which does not obviously appear in the mechanics leading to the derivation of that equation. For example, the dynamics of the collisions we consider are truly reversible and obey conservation of momentum and energy, yet the consequence of such collisions is to lead uniquely to the Maxwellian distribution, at least in the sense that the latter is overwhelmingly more probable than any other distribution, once equilibrium is established. The details of the controversy between Boltzmann and some of his contemporaries can be studied in the excellent selections of original papers edited by Brush[3]; here, we wish to comment on some aspects of irreversibility as it now appears.

Before going on, it is worth making the obvious point that nature is rarely, if ever, truly reversible. No heat engine can work on a Carnot cycle; no collision really conserves energy; no electron motion is uncoupled to electro-magnetic fields; and so on. So, in a very real sense, the argument about reversibility is not sensible and Boltzmann's critics would have been well advised to welcome the fact that his theory did lead to irreversible phenomena. Their point should be paraphrased by asking how it is that reversible *assumptions* lead to irreversible results. We have, of course, said earlier that *irreversible* assumptions, such as the Langevin theory, lead to irreversible results and therefore, presumably, nobody would criticize Langevin on this score. One must assume, then, that if Boltzmann had calculated, for example, a collision term of the Fokker–Planck type, which he might have done because Markov's early work could have been available to him, his work would have been unobjectionable.* This turns out to be a rather useful line of thought.

It has been pointed out in several places that the discussion of the collision dynamics as carried out by Boltzmann is to some extent incomplete. Briefly, Boltzmann assumes that two particles are known to be located in such a way that they are on a collision course. The collision period starts when they are some considerable distance apart and continues until they have separated by about the same distance, the actual time that they are in near contact being explicitly assumed to be a small proportion of this total time. The question which remains unasked by Boltzmann is how they arrived at their known starting conditions.

Looking at this problem from the point of view of present-day probability theory, we should say that they arrive as a result of a chain of Markov processes, if we want the simplest probabilistic description. However, if we attempt to look at the motion of the particles over a few collisions, we find that the results of assuming an initial state and working forward to the collision state are not even given by the same equations as for the converse of working back from after the collision. In probability theory, these are called the forward and backward Chapman–Kolmogorov equations, but the point here is merely that they are different. Without any detail, we can say that, although the colliding pair may be thought of as being 'random', they have also reached their initial state by following the chain of events which had the highest probability. This is perhaps more evident if we think of the evolution of a large sub-system of particles. Therefore, the Boltzmann collision correctly represents the deterministic effect of a collision, but disguises the fact that the collisions are part of a probabilistic sequence. If this is the case, it is hardly surprising that collisions in a large population with an initially specified distribution will drive it towards the most probable distribution.

It therefore seems that it is not necessary to invoke dissipation to explain irreversibility: large ensembles of non-dissipative systems would show irreversible behaviour. On the other hand, it seems clear that dissipation due perhaps to (neglected) coupling mechanisms will speed up the attainment of equilibrium in real experiments.

*The dates are: Boltzmann 1844–1906, Markov 1856–1922.

8

Plasma Kinetics

In view of the rapid development which has taken place in plasma kinetic theory since intensive work on thermonuclear fusion was begun, in the nineteen fifties, it will not be possible to do more than introduce the subject in this book. However, both because of the interest in the theory and because of the high importance of fusion research towards a long-term solution of the world's energy problem, it is worthwhile to discuss a few of the more important ideas. We introduce the subject by discussing some of the fundamental quantities which characterize a plasma.

8.1 Plasma characteristics

The term 'plasma', introduced by Langmuir, is used to mean a region filled with ionized gas. The manner of ionizing the gas is inessential, but typical methods are by passing an intense electric discharge or by raising the gas to a very high temperature. The plasma is said to be fully ionized when all the gas molecules in the volume have been reduced to positive ions plus electrons. If neutral molecules are still present, the plasma is partially ionized, a more complicated state of affairs and one which is not much treated in plasma theory. Let us now suppose that the plasma was formed from a single gas and that only one species of positive ion is present, charge q. Generally, of course, $q = |e|$, but multiple ionization is possible.

The plasma in equilibrium is electrically neutral, so that

$$\langle N_\mathrm{p}\rangle q + \langle N_\mathrm{e}\rangle e = 0 \tag{8.1}$$

i.e., when $q = |e|$, $\quad \langle N_\mathrm{p}\rangle = \langle N_\mathrm{e}\rangle$ (8.2)

The average distance between ions (we shall soon see that the ions are, relatively speaking, stationary in space) is approximately

$$l = 1/\langle N_\mathrm{p}\rangle^{1/3} \tag{8.3}$$

More accurately, the most probable value of l is $0.554\,\langle N_\mathrm{p}\rangle^{-1/3}$.

We shall now show that, although large volumes of plasma are neutral, there are small-scale spatial variations in the electrical potential. The physical basis of this is readily understood by the following argument. If we take an origin at the centre of an ion, the coulomb force between the ions and neighbouring electrons attracts electrons nearer to the ion than they would otherwise be. Thus, each ion may be thought to be associated with a cloud of electrons such that the density of the cloud decays radially as one moves away from the ion. Clearly, this concept is a little aphysical in that it is obviously necessary for the total charge of the cloud to be equal in magnitude to the ion charge, so that the cloud is one electron. As the theory develops, we shall find that we really have to think in terms of a sphere of determinate radius which contains many ions and therefore many electrons, and the potential is expressed with respect to the centre of this sphere. This concept makes sense only if the coulomb energy is small compared with the thermal energy, because otherwise the plasma behaviour would be determined entirely by the long-range coulomb collisions. Numerically, for the development to apply, we must then have,

$$e^3/\epsilon_0 l \ll kT \tag{8.4}$$

Consider the spherical volume centred on an ion. Let the potential be $\phi(r)$ and take a small volume element centered at radius r. Then, using the Boltzmann theorem, the numbers of ions and electrons in $\mathrm{d}V$ are given by

$$N_\mathrm{e}\mathrm{d}V = \langle N_\mathrm{e}\rangle \exp(+e\phi/kT)\,\mathrm{d}V$$

$$N_\mathrm{p}\mathrm{d}V = \langle N_\mathrm{p}\rangle \exp(-q\phi/kT)\,\mathrm{d}V$$

since the electron density is locally increased while the ion density is decreased. Since $e\phi/kT$ is small, we can expand the exponentials and, using equation 8.1, obtain an expression for the total charge density in $\mathrm{d}V$, viz. $eN_\mathrm{e} - qN_\mathrm{p}$, which is

$$\rho(r) = -\frac{e^2}{kT}(\langle N_\mathrm{e}\rangle + \langle N_\mathrm{p}\rangle)\,\phi(r)$$

Poisson's equation then gives

$$\nabla^2\phi(r) = -\rho(r)/\epsilon_0 = \kappa^2\phi(r) \tag{8.5}$$

with $\kappa^2 = e^2(\langle N_\mathrm{e}\rangle + \langle N_\mathrm{p}\rangle)/\epsilon_0 kT$ 8.6

$$= 2e^2\langle N_\mathrm{e}\rangle/\epsilon_0 kT$$

The solution of equation 8.5, which behaves like the Coulomb potential $(-e/\epsilon_0 r)$ for small r, is

$$\phi(r) = -e\exp(-\kappa r)/\epsilon_0 r \tag{8.7}$$

which clearly falls off considerably more rapidly than the Coulomb potential for $\kappa r > 1$.

The Debye length is defined as κ^{-1}, viz.

$$\lambda_\mathrm{D} = \left(\frac{\epsilon_0 kT}{2e^2\langle N_\mathrm{e}\rangle}\right)^{1/2} \tag{8.8}$$

The number of particles in a sphere whose radius is λ_D is $(8\pi\lambda_\mathrm{D}^3/3)\,\langle N_\mathrm{e}\rangle$. Using equations 8.8 and 8.4, this leads to the condition $\langle N_\mathrm{e}\rangle\lambda_\mathrm{D}^3 \gg 8\pi/3 \gg 1$, or there must be a large number of particles in the Debye sphere if the present theory is valid. Numerically, the Debye length is given by $\lambda_\mathrm{D} = 69.0\,(T/N_\mathrm{e})^{1/2}$ metres, and some values are shown in Table 1.

With an understanding of the Debye length we can begin to understand why the behaviour of a plasma is considerably more complicated than that of a cloud of electrons or ions. If we imagine that a low-frequency electric field is applied to a region containing plasma, the ions will respond to the field variations and will, so to speak, drag the much more mobile electrons along with them. Thus, the plasma as a whole is influenced by l.f. electric fields. Equally, a magnetic field deflects the plasma as a whole rather than separating out the ions and electrons. These ideas suggested to Langmuir a jelly-like behaviour, and hence the name. Another important fact about the Debye length is that if the plasma is in contact with a bounding surface, for example in a long laboratory discharge tube, free charges can be formed

TABLE 1

Debye length in metres for a few typical plasmas. The lengths are orders of magnitude only.

	T	N_e	λ_D
Ionosphere	2×10^2	10^5	3.0
Solar corona	2×10^6	10^6	100.0
Glow discharge	10^5	10^{10}	0.2
Thermonuclear fusion reactor	10^8	10^{14}	7×10^{-2}
MHD device	10^4	10^{16}	7×10^{-5}
Solids	300	10^{22}	10^{-8}

on the walls, and the discontinuity in potential in moving from the wall to the plasma will be of the order of a Debye length, since the effect of the free charges will be screened out over that distance.

Another important characteristic number for a plasma is the plasma frequency. This can be defined either for electrons or for ions, but the term 'plasma frequency' when unqualified usually means the electron plasma frequency, symbol ω_{pe}. The definition is

$$\omega_{pe}^2 = \frac{e^2 \langle N_e \rangle}{\epsilon_0 m} \quad 8.9(a)$$

Another relationship for ω_{pe}^2 is

$$\omega_{pe}^2 = \frac{kT}{2m\lambda_D^2} \quad 8.9(b)$$

The corresponding ion plasma frequency is

$$\omega_{pi}^2 = \frac{q^2 \langle N_p \rangle}{\epsilon_0 M_i} \quad 8.10$$

clearly two or three orders of magnitude *smaller* than ω_{pe}. The simplest physical meaning of ω_{pe} is that, if we create an excess of electron charge in some region of a plasma by, for example, pulsing a probe positively for a short time, the excess charge will redistribute itself in oscillations at frequency $\omega_{pe}/2\pi$ once the disturbing potential is removed.

Another property depending on the plasma frequency is the transmission of electromagnetic waves. If we try to transmit e.m. radiation through a large, plane plasma system, we find that transmission is impossible for incident frequencies *below* the plasma frequency, because the charges have sufficient time to redistribute themselves so as to screen out the fields at lower frequencies. At frequencies above ω_{pe} the electrons cannot follow the waves, and some transmission results. At much higher frequencies the plasma is fully transparent to e.m. radiation. We discuss this more fully below.

8.2 The scope of plasma kinetic theories

Before entering into details, it seems desirable to describe briefly the scope of plasma kinetic theories, as there is some confusion in the literature about what they ought to accomplish. First, kinetic theory is obviously concerned with the development in time of the probability distribution functions relating to the electrons, various species of ions, etc. The situations studied are often far away from any conceivable equilibrium state. How, then, was the plasma prepared and how are any electromagnetic fields relevant to particular problems coupled to the plasma? In a great deal of work on plasma kinetics these points are left unclear, and confusion has resulted. The problem can be made concrete by considering the way in which the theory of microwave devices using long electron beams is set up. As example, think of a travelling-wave tube. This has an intense electron beam of energy between a few hundred and many thousand eV, a slow-wave circuit which propagates e.m. waves whose longitudinal phase velocity is below that of light and, by appropriate design, can be made to equal the average electron velocity for some chosen energy. Waves induced in the circuit by a high-frequency generator interact with the electrons so that, if the electron velocity is near the synchronous value, energy is either taken by the circuit from the electrons, for an electron velocity very slightly above the synchronous value, or, when the velocity is below the synchronous value, the circuit transfers power to the beam. If the electrons are *exactly* equal in velocity to the phase velocity, there is no interaction.

The theory is set up by deriving two equations: the so-called 'circuit' equation, which gives the self-consistent fields on the circuit when a particular electron beam excitation is applied as driving term, and the so-called 'electronic' equation which gives the electron current caused by an applied electric field. These two relationships must apply simultaneously to the problem. The resulting equation, called the 'determinantal' equation, can be solved for the propagating wave-functions, of which there are four in the travelling-wave tube. These wave functions in general have large imaginary parts and small real parts, and the wave which is utilized to give power amplification is one whose amplitude increases with distance from the input end of the tube. The travelling-wave tube is inherently unstable and, if no avoiding action is taken, will oscillate if the beam current or tube length is increased so that the gain exceeds a few decibels.

Of course, the electron beam is not at all like a plasma. The velocity distribution function is very narrow, a small fraction of an eV in (at least) several hundred; therefore, for the basic theory, one can treat it as delta function and the electron behaviour can be calculated by deterministic mechanics without bothering about the transport equations. However, the transport equation can be used and has been used to study questions such as what happens when the beam distribution function is made broad? Thus, kinetic theories *can* be used to formulate the electronic equation, although there is frequently no need for them. The point is, however, that the most elaborate kinetic analysis alone cannot contain all the required information. Without the circuit equation, kinetic theory will allow calculation of the particulate distribution functions in response to assumed fields or on relaxation after fields are removed, but it clearly can not do more. This is not, however, to say that kinetic theories do not supply enough information for the solution of properly set problems: the worries start when their range is extended beyond their proper province.

In conclusion, a few words on plasma microwave amplifiers and oscillators may be helpful. Several types of such devices have been described. Usually, an electron beam is fired through a plasma region, and several detailed mechanisms for amplification are known. A simple example is that of a long glass tube filled with plasma, an electron beam being focused along the axis. In this case, the circuit equation alone shows that the tube filled with plasma in external free space will propagate what is called a 'slow backward wave'. This means that the phase velocity is $<c$ and that the phase and group velocities are in opposite senses. If the beam carries large enough current, the device oscillates. The mechanism is that the electrons interact with the forward (phase velocity) circuit wave, but the reverse group velocity transfers energy back to the input end of the device, giving rise to a backward wave oscillation. In this case, and in several other tubes, the plasma should be regarded as part of the circuit, since it forms a region of altered dielectric constant which modifies the circuit equations in a desirable manner. It should be remembered that very few plasma electrons in a laboratory plasma have velocities near that of the beam electrons, so the division into plasma plus beam is not purely artificial.

8.3 The Vlasov equation

Turning now to the application of kinetic theory to plasma processes, we usually implicitly consider an infinite plane plasma system. Electromagnetic waves are assumed to have been coupled to the plasma at some point to the left of the space origin or, alternatively, it is assumed that the plasma, having been excited, is allowed to relax. Thus, in either case there is no e.m. wave-propagating structure *inside* the plasma region considered.

In Chapter 7, we introduced the Vlasov equation[1] as the Boltzmann equation with the right-hand side reduced to zero. We must now refine this idea somewhat and also specifically consider the meaning which might be ascribed to the force terms on the l.h.s. of the equation.

First, why can we, in a certain approximation, disregard the collision term on the r.h.s. of equation 7.4? Clearly, the collisions are not absent, nor is the plasma in equilibrium, so that one might rationally suppose that the collision integral is zero. Instead, what is here assumed is that the binary or particle–particle collisions may be neglected in comparison with the long-range interaction due to the Coulomb force, whether shielded or (even more so) unshielded. These forces are, of course, the space-charge forces of electron beam theory, but in a plasma the condition of average neutrality restricts their range to distances of the order of the Debye length, instead of restrictions due to the dimensions of the apparatus. Moreover, we have already said that the Coulomb interaction is fundamental in supplying the long-range cohesion required for plasma-like behaviour. Thus, if we put the Coulomb forces on the l.h.s., we can take into account the long-range part of the collisions in which many particles interact with one another, neglecting the binary collisions on the r.h.s. In other words, we have to use self-consistent fields. However, we must also recognize that the plasma will, after a fairly long time, revert to a state which represents equilibrium. Then, the binary collisions can no longer be neglected, since they are the mechanism for maintaining long-term equilibrium. Therefore, the Vlasov equation is characterized (a) by the use of self-consistent fields and (b) by the fact that it applies only for times less than the relaxation time.

Now, we can rewrite equation 7.4 for particles of species α and in velocity space as

$$\frac{\partial f_\alpha}{\partial t}+(\mathrm{v}\,.\,\nabla)f_\alpha+e_\alpha\left\{E+\frac{1}{c}(\mathrm{v}\times H)\right\}\nabla \mathrm{v} f_\alpha=0 \qquad 8.11$$

where ∇f_α is the ordinary divergence and $\nabla \mathrm{v} f_\alpha$ is the divergence with respect to velocities. $\boldsymbol{E}$ and $\boldsymbol{H}$ satisfy Maxwell's equations *including* currents and free charges, which are themselves defined through

$$\rho=\sum_\alpha e_\alpha\int f_\alpha(r,\mathrm{v},t)\,\mathrm{d}V \qquad 8.12$$

$$j=\sum_\alpha e_\alpha\int \mathrm{v} f_\alpha(r,\mathrm{v},t)\,\mathrm{d}V \qquad 8.13$$

These last two equations stress the self-consistent nature of equation 8.11, because f determines ρ and j which, in turn, help to determine $\boldsymbol{E}$ and $\boldsymbol{H}$. The general technique for the solution is that of perturbation. Some reasonable distribution function is assumed for f_0, which often departs very considerably from a Maxwellian, and a solution is sought for $f=f_0+f_1$, where f_1 is a time-dependent perturbation. If we consider that f_0 is independent of spatial coordinates and time but is a function of v, which would of course be true for a Maxwellian f_0, and specialize equation 8.11 for electrons, we obtain

$$\frac{\partial f_1}{\partial t}+\mathrm{v}\,.\,\frac{\partial f_1}{\partial r}+\frac{e}{m}\left\{E+\frac{1}{c}(\mathrm{v}\times H)\right\}\frac{\partial f_0}{\partial \mathrm{v}}=0 \qquad 8.14$$

Equation 8.14 applies for fairly high frequencies, at which it is permissible to consider the ions to be stationary. It is now possible to solve equation 8.14 for f_1 in terms of $\boldsymbol{E}$ and $\boldsymbol{H}$ and thus achieve a first-order solution of our problem.

There are many techniques for the solution of equation 8.14 and, before solving it, it will be useful to the reader to know that the solution has caused violent controversy, which is reflected in the literature to such a large extent that it cannot be ignored, although in retrospect it is rather sterile. Vlasov's original solution[1] gave a result in which undamped waves, of frequency very near to the plasma frequency, could propagate in the plasma. Landau[2] severely criticized this result and, using Fourier–Laplace transform techniques, gave a result which included Vlasov's value for the propagation constant, but with an additional damping term.

Although, from what has been said in the introductory remarks, it would seem physically necessary for the fields eventually to decay back to an equilibrium, many writers appear to have thought that Landau damping was introduced by his mathematics rather than by the real physics of the problem. Numerous methods of solving the equation by techniques not involving Laplace transforms were therefore published and, when mathematically correct, showed that the damping really appeared in the solution. The basis of the controversy seems really to have resided in lack of clarity in formulating the initial problem without well-prescribed initial and boundary conditions. Had this been done, there should have been no controversy, since, physically, Landau damping is simply the interaction phenomenon observed with pure electron beams of which we have spoken. In the plasma case,

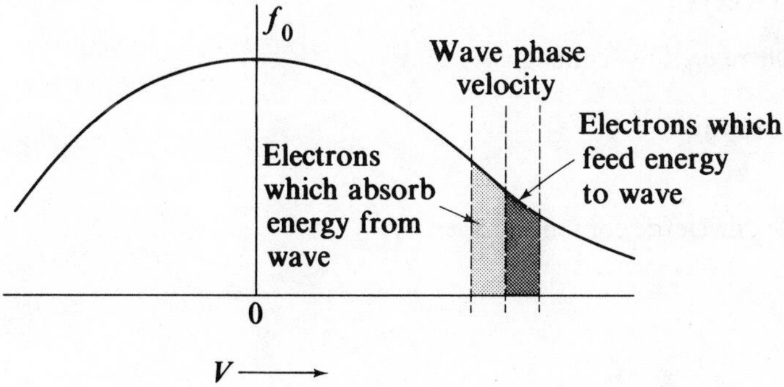

Fig. 8.1 The Landau damping phenomenon for a Maxwellian particle energy distribution.

the electron velocity distribution function is very broad (Fig. 8.1) and, at a given frequency, there are fewer electrons travelling with a velocity just above the wave phase velocity than just below this velocity. We have already said that electrons with velocities above the phase velocity give energy to the wave, and vice versa; therefore, a small overall damping of the waves results from a broad velocity distribution function of symmetrical but not necessarily Maxwellian shape. Electrons which depart from the wave velocity by more than a (small) determinable value are extremely weakly coupled to the wave and their influence is negligible.

To summarize this discussion, we can say that there are several techniques for solving equation 8.14, all of which can be applied in several degrees of approximation, but the closer approximations yield solutions which correspond with propagating waves damped in the time coordinate.

Turning now to the solution of equation 8.14, we find that we can simplify it for distribution functions f_0 which are *isotropic,* for then $\partial f_0/\partial \mathbf{v} = \mathbf{v} \,.\, \partial f_0/\partial(\mathbf{v})^2$ and

$(\mathbf{v} \times \boldsymbol{H}) \,.\, v$ is clearly zero. since $\mathbf{v} \times \boldsymbol{H}$ is perpendicular to both $\mathbf{v}$ and $\boldsymbol{H}$. Thus, we must solve

$$\frac{\partial f_1}{\partial t} + \mathbf{v} \,.\, \frac{\partial f_1}{\partial r} + \frac{e}{m} \boldsymbol{E} \,.\, \frac{\partial f_0}{\partial \mathbf{v}} = 0 \qquad 8.15$$

Here, we solve equation 8.15 by a substitution method, letting $f = f_1(0)$ $\exp \mathrm{j}(\omega t - \boldsymbol{k} \,.\, \boldsymbol{r})$, a wave propagating along the positive r direction with phase velocity $p = \omega/|k|$. The substitution is equivalent to using a Fourier transform in either space or time followed by a Laplace transformation in either time or space. Since $\boldsymbol{E}$ is a function of f_1, through equation 8.12 and Maxwell's equation $\epsilon_0 \operatorname{div} \boldsymbol{E} = \rho$, $\boldsymbol{E}$ must have the same wave function as f_1, and we find

$$f_1 = -\frac{\mathrm{j}(e/m)\boldsymbol{E}\, \partial f_0/\partial \mathbf{v}}{(\omega - k\mathbf{v})} \qquad 8.16$$

Therefore, if we let J = electric current density,

$$J = -j\frac{e^2}{m} E \int \frac{\mathbf{v}(\partial f_0/\partial \mathbf{v})\, \mathrm{d}\mathbf{v}}{(\omega - k\mathbf{v})} \qquad 8.17$$

Since $J = \sigma_{\mathrm{ij}} E$, where σ_{ij} is the conductivity tensor

$$\sigma_{\mathrm{ij}} = -j\frac{e^2}{m} \int \frac{\mathbf{v}(\partial f_0/\partial \mathbf{v})\, \mathrm{d}\mathbf{v}}{(\omega - k\mathbf{v})} \qquad 8.18$$

and since the complex dielectric constant is given by

$$\epsilon_{\mathrm{ij}} = \delta_{\mathrm{ij}} + j\frac{\sigma_{\mathrm{ij}}}{\epsilon_0 \omega} \qquad 8.19$$

then

$$\epsilon = 1 + \frac{e^2}{\epsilon_0 m \omega} \int \frac{\mathbf{v}(\partial f_0/\partial \mathbf{v})\, \mathrm{d}\mathbf{v}}{(\omega - k\mathbf{v})} \qquad 8.20\text{(a)}$$

$$= 1 - \frac{e^2}{\epsilon_0 m \omega k} \int \frac{\mathbf{v}(\partial f_0/\partial \mathbf{v})\, \mathrm{d}\mathbf{v}}{(\mathbf{v} - \omega/k)} \qquad 8.20\text{(b)}$$

Equation 8.20(b) highlights the source of the difficulty of the final solution. The integrand has a pole at $\mathbf{v} = \omega/k$, that is it is singular when the wave velocity is synchronous with the velocity v. As a first step towards carrying out the integration in equation 8.21, let us tidy up the notation and transform the integral by integrating by parts.

First, in equation 8.21 f_0 is normalized to N_{e} and not to unity. Let us take the N_{e} outside the integral and insert $\omega_{\mathrm{p}}{}^2 = e^2 N_{\mathrm{e}}/\epsilon_0 m$. Then, partial integration gives

$$\int \frac{\mathbf{v}(\partial f_0/\partial \mathbf{v})\mathrm{d}\mathbf{v}}{(\omega - k\mathbf{v})} = -\omega \int \frac{f_0\, \mathrm{d}v}{(\omega - k\mathbf{v})^2}$$

because $f_0(v)$ must vanish as $v \to \pm\infty$. Then equation 8.20 becomes

$$\epsilon = 1 - \frac{\omega_p^2}{\omega^2}\int \frac{f_0\,dv}{(1 - kv/\omega)^2} \qquad 8.21$$

Equation 8.21 is the final general result for the complex permittivity. The dispersion equation, the relation between ω and $\boldsymbol{k}$, is obtained by putting $\epsilon = 0$ or

$$\omega^2 = \omega_p^2 \int \frac{f_0\,dv}{(1 - kv/\omega)^2} \qquad 8.22$$

We can evaluate the integral for various assumed equilibrium distributions. The most common example is a Maxwellian f_0, viz.

$$f_0(v) = \left(\frac{m\beta}{2\pi}\right)^{3/2} \exp\left(-\frac{m\beta v^2}{2}\right), \quad v^2 = v_x^2 + v_y^2 + v_z^2$$

or, for one velocity component,

$$f_0(v_z) = \left(\frac{m\beta}{2\pi}\right)^{1/2} \exp\left(-\frac{m\beta^2 v_z^2}{2}\right)$$

If we use the latter to derive ϵ_z, we get the dispersion equation for longitudinal waves in the form

$$\omega^2 = \omega_p^2 \left(\frac{m\beta}{2\pi}\right)^{1/2} \int_{-\infty}^{+\infty} dv_z \left(1 - \frac{kv_z}{\omega}\right)^{-2} \exp\left(-\frac{m\beta v_z^2}{2}\right)$$

$$\approx \omega_p^2 \left(\frac{m\beta}{2\pi}\right)^{1/2} \int dv_z \left(1 + \frac{2kv_z}{\omega} + \frac{3k^2 v_z^2}{\omega^2} + \ldots\right) \exp\left(-\frac{m\beta v_z^2}{2}\right)$$

$$= \omega_p^2 \left(1 + \frac{3k^2}{m\beta\omega^2} + \ldots\right) \qquad 8.23$$

Since terms which are odd in v_z vanish, equation 8.23 explicitly neglects terms in k^4/ω^4 or in p^{-4}. Since p is a large number O(c), the accuracy is good.

Eliminating β by using λ_D* we reach

$$\omega^2 \approx \omega_p^2 \left(1 + 3\lambda_D^2 k^2 \frac{\omega_p^2}{\omega^2} + \ldots\right) \qquad 8.24$$

Equation 8.24 is usually quoted in the literature in the form

$$\omega = \omega_p(1 + 3\lambda_D^2 k^2/2) \qquad 8.25$$

which is valid for $\omega_p \approx \omega$, which (see Fig. 8.2) automatically ensures that $3\lambda_D^2 k^2/2 \ll 1$. Equation 8.25 is often called the 'Bohm–Gross dispersion relation'.[3]

*Since the derivation tacitly assumes the *ions* are at rest, in this case $\lambda_D^2 = \epsilon_0 kT_e/e^2 N_e$, where T_e = electron temperature.

However, it is clear that the integration of equation 8.22 is not really satisfactory. The denominator of equation 8.22 gives rise to a pole at $\omega = kv$ and we have not considered modifications which ought to be made to allow ω to become complex. If we write $\omega = \omega' + j\omega''$ in the wave functions, we get waves which decay in time and are therefore the expected result of the analysis. The denominator in, for example, equation 8.20 then becomes $(\omega' - k\mathrm{v}) - j\omega''$, and subsequent integrals have to be evaluated with this understanding.

Using contour integration,* we can see that a contribution of $j\pi$ x residue at $\mathrm{v} = \omega/k$ should be added to the r.h.s. of equation 8.23. Notice that the residue depends on $(\partial f_0/\partial \mathrm{v})_{v\,=\,\omega/k}$, that is on the slope of the distribution function at $\mathrm{v} = \omega/k$, which is consistent with our understanding of Landau damping as caused by the difference in the number of electrons with velocities just above and just below the synchronous value. Carrying out the details for the Maxwellian f_0, we find that equation 8.24 should be modified to

$$\omega^2 = \omega_p^2\left(1 + 3\lambda_D^2k^2\frac{\omega_p^2}{\omega^2} - j\sqrt{\left(\frac{\pi}{8}\right)}\cdot\left(\frac{\omega}{\omega_p}\right)^3\cdot\frac{2}{(\lambda_Dk)^3}\exp\left[-\left(\frac{\omega}{\omega_p}\right)^2\cdot\frac{1}{2(\lambda_Dk)^2}\right]\right)$$

Therefore

$$\omega \approx \omega_p\left(1 + \frac{3}{2}\lambda_D^2k^2\frac{\omega_p^2}{\omega^2} - j\sqrt{\left(\frac{\pi}{8}\right)}\left(\frac{\omega}{\omega_p}\right)^3\cdot\frac{1}{(\lambda_Dk)^3}\exp\left[-\left(\frac{\omega}{\omega_p}\right)^2\cdot\frac{1}{2(\lambda_Dk)^2}\right]\right)$$

or $$\omega' \approx \omega_p(1 + 3\lambda_D^2k^2/2) \qquad 8.26$$

$$\omega'' = -\omega_p\sqrt{\left(\frac{\pi}{8}\right)}\frac{1}{(\lambda_Dk)^3}\exp\left[-\left(\frac{1}{2\lambda_Dk^2} + \frac{3}{2}\right)\right] \qquad 8.27$$

Therefore, equation 8.25 is actually correct for $\mathrm{Re}(\omega)$, while equation 8.27 gives the Landau damping term, for the special case of Maxwellian f_0.

In plasma physics, the problem of distributions f_0 which might give rise to *growing* longitudinal oscillations is extremely important, since instabilities of this kind may contribute to the stability problems which have, so far, prevented the achievement of long-period plasma confinement. The literature on these topics is excellently summarized in several textbooks: the treatments due to Ecker[6] and by Krall and Trivelpiece[7] are especially recommended.

For *transverse* waves we can be brief. It is a standard result of electromagnetic theory that k^2 has to be replaced by $k^2 - \omega^2/c^2$. The waves are hardly coupled to the plasma, which simply appears as a medium of permittivity $\epsilon_0(1 - \omega_p^2/\omega^2)$. The dispersion relationship is

$$\omega^2 = \omega_p^2 + k^2c^2 \qquad 8.28$$

*This integration is treated at great length in the literature on Landau damping, but it is standard mathematics.[4,5]

8.4 Discussion of Landau damping

8.4.1 Longitudinal waves

Fig. 8.2 shows a plot of the real part of ω versus the propagation constant k. Such plots are called 'Brillouin diagrams'. Since the relationship is $\omega = \omega_p(1 + 3k^2\lambda_D^2/2)$, $\omega = \omega_p$ when $k = 0$ and the curves are parabolic. Two cases, for identical values of ω_p but different values of λ_D are shown. Physically, this means we have compared two plasmas which have identical densities of particles, but that denoted by λ_2 corresponds with a higher electron temperature than the other.

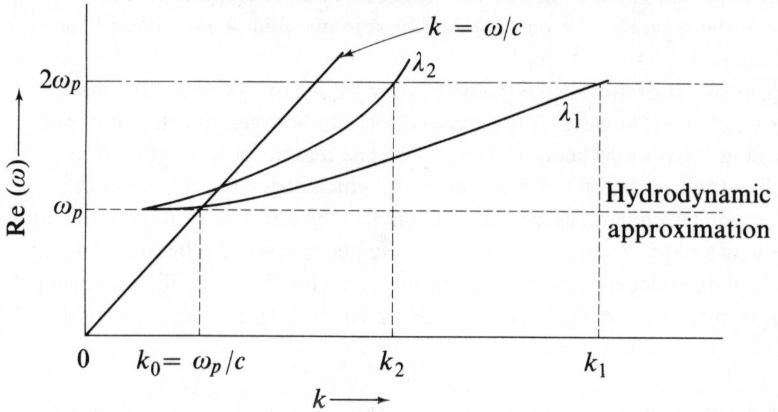

Fig. 8.2 Brillouin ($\omega \sim k$) diagram for a warm plasma.

The hydrodynamic approximation which applies for a *cold* plasma, $T \to 0$, is a horizontal straight line. The straight line $k = \omega/c$ divides the diagram into a left-hand region, in which the phase velocity of the waves is always *greater* than c, and a right-hand region in which the phase velocity is *less* than c. This line intersects the horizontal ω_p at $k_0 = \omega_p/c$ and intersects the parabolas for λ_1 and λ_2 quite near to this point. The reader may recall that the Brillouin diagram for an air-filled wave-guide is a parabola starting at the cut-off ω_c which lies entirely to the left of $k = \omega/c$ and only approaches asympotically to this value as $k \to \infty$. He will also recall that waves whose angular frequency is *below* ω_c will *not* propagate but, instead, will exist as evanescent waves around any change in waveguide dimensions or obstacles. In the plasma, the waves may have phase velocities greater than c, if their frequency is only just above ω_p, but for higher frequencies the phase velocity drops below c. These waves all propagate, whether they are fast or slow, but the waves in the range where $p > c$ are nearly *undamped* because, according to the relativistic increase in mass, there are no plasma electrons with velocities higher than c and the coupling between the wave and the electrons is negligible. Thus, disturbances at frequencies very little above ω_p propagate without Landau damping, although, of course, they may be subject to collisional damping, which is not included in the Vlasov theory.

For higher frequencies, the damping coefficient is given by

$$\mathrm{Im}(\omega) = -\sqrt{\left(\frac{\pi}{8}\right)}\frac{\omega_p}{|\lambda_D{}^3k^3|}\exp\left[-\left(\frac{1}{2\lambda_D{}^2k^2}+\frac{3}{2}\right)\right]$$

This damping is in time, so that the wave functions for fields, potentials, etc. are of the form

$$E(z, t) = E_0 \exp \mathrm{j}\,[kz - \omega t] \ \exp \mathrm{Im}(\omega) t$$

This expression is invalid both for very small k and for very large k. At small k, a relativistic correction is required, while when $k\lambda_D$ is O(1.0) the wavelength is of the same order as the radius of the Debye sphere and the shielding concept breaks down. Waves at frequencies below ω_p will not propagate and are evanescent waves localized in the region of the wave coupler or antenna used in any attempt to excite them.

Experiments made to test the theory[7,12] give excellent agreement for the damping coefficient; however, the measured relation between ω and k does not agree at all well with the theoretical parabola. The reason for this is that the measurements are made on a plasma geometry which differs sharply from the infinite planar system and, as we have said earlier, the dispersion or circuit equation for a finite geometry departs radically from the planar case. On the other hand, since the damping depends essentially on the interaction between the waves and particles, it might be expected that it would be less influenced by geometrical factors.

8.4.2 Transverse waves

The dispersion equation in this case is $\omega^2 = \omega_p{}^2 + k^2c^2$. Then, the phase velocity $\rho = c(1 + \omega_p{}^2/c^2k^2) > c$ is invariable. For the reason cited above, the damping is negligible.

8.5 The relaxation model in plasmas

It is now reasonable to investigate the application to plasmas of the simple relaxation time approximation $(\partial f/\partial t)_c = (f_0 - f)/\tau_e$ described in section 7.3.*

Such a collision term would allow, at least in part, the inclusion of effects due to binary collisions. These effects must be investigated for the following reasons. The Vlasov equation allows the derivation of expressions for the electric conductivity produced by the fields set up by inhomogeneities in the plasma. However, the plasma conductivity is high and the self-consistent fields are very low; therefore, the drift velocities due to the perturbations are small. On the other hand, collisions give rise to the process of diffusion, and the diffusion current should certainly be evaluated so as to determine its order of magnitude relative to the drift current. Exactly the same kind of comment can be made on mass-transport coefficients, heat and energy transport.

In Chapter 7, we have already derived an expression, equation 7.26, which gives the longitudinal component of the conductivity or, strictly, that part of the con-

*In plasma physics, this is often called the 'Bhatnagar–Gross–Krook technique', BGK for short.

ductivity which is due to the electrons. For the plasma electrons, this is

$$\sigma(\omega) = \frac{N_e e^2}{m}\left[\frac{1 - j\omega\tau_e}{1 + (\omega\tau_e)^2}\right] \qquad 8.29$$

Equation 8.29 was derived for no magnetic field. It is left as an exercise for the reader to derive the result for a magnetic field B in the direction of propagation.

Exercise
Show, for harmonically varying fields in which the direction of propagation is along the magnetic field vector, that

$$[\sigma_{ij}] = \frac{N_e e^2}{m}\begin{bmatrix} \dfrac{\nu_c + j\omega}{(\nu_c + j\omega)^2 + \omega_c^2} & \dfrac{-\omega_c}{(\nu_c + j\omega)^2 + \omega_c^2} & 0 \\ \dfrac{\omega_c}{(\nu_c + j\omega)^2 + \omega_c^2} & \dfrac{\nu_c + j\omega}{(\nu_c + j\omega)^2 + \omega_c^2} & 0 \\ 0 & 0 & \dfrac{1}{\nu_c + j\omega} \end{bmatrix} \qquad 8.30$$

where $\omega_c^2 = e^2 B/m$, i.e. ω_c is the cyclotron frequency, and $\nu_c = 1/\tau_e$. σ_{12} and σ_{21} are the Hall coefficients for the plasma.

It is not necessary for the purposes of this book to develop further transport coefficients; we are interested only in the method.

We now notice that equation 8.30 is useless until we can assign a value to τ_e or ν_c. This problem is difficult, both in principle and in mathematical detail. The difficulty in principle is largely due to the long-range nature of the coulomb forces. In Appendix 6 we discuss the Rutherford scattering formula, equations A6.14 and A6.15. We shall find that attempts to integrate over all values of the impact parameter lead to divergence at the limits, and special assumptions have to be made to overcome this.

Calculations of the relaxation time are based on studies of the behaviour of a test particle, of fixed charge e and mass m, and known initial velocity v whose motion in the plasma is traced.* The plasma particles are called the 'field' particles and their number density is N_s, charge e_s and velocity v_s. The subscript denotes that these are the scattering particles. We already know, from appendix 6, that a single scattering event turns the relative velocity vector $u = \mathbf{v} - \mathbf{v}_s$ through a small angle θ, which depends on the impact parameter and the energy according to equation A6.14, Appendix 6. Then, the test particle in time dt is subject to $N_s u\, dt$ scattering events and the probability of being scattered into the solid angle $d\Omega = 2\pi \sin\theta d\theta$ is $N_s u\sigma(\theta_1 u)d\Omega dt$ from the manner in which $\sigma(\theta)$ was defined in the appendix.

The following difficulty now arises. From Appendix 6, $\sigma(\theta, u)$ can be written as

*Strictly, the test particle is one of a large number of identical test particles, all with the same relative velocity, etc. This allows initial averaging so that collisions in one plane are considered.

$R \sin^{-4}(\theta/2)$ and, using the expression for $\Delta \boldsymbol{p}$, we see that averaging over θ involves the integral

$$\int_0^\pi \frac{\sin^2(\theta/2) \,.\, \sin\theta \, d\theta}{\sin^4(\theta/2)} = \int_0^\pi \cot(\theta/2)\, d\theta$$

$$= \Big[-\ln \sin(\theta/2) \Big]_0^\pi = \Big[\ln \frac{1}{\sin(\theta/2)} \Big]_0^\pi \qquad 8.31$$

The last form is clearly infinite at the lower limit and this fact, in turn, comes back to the long-range nature of the coulomb force. This difficulty is handled in the following manner. It is assumed that the plasma characteristics are such that there are many electrons in a Debye sphere. Then, the Debye sphere is thought of as a kind of quasi-particle which takes part in the collisions with the test particles throughout their path between collisions. However, for impact parameters $b > \lambda_D$, since the shielded coulomb potential has fallen off sharply, well below the unshielded value, the collision cross-section drops to zero. For the minimum value of b it is usual to take the special value $b_\perp$, which gives $\theta = \pi/2$. Such close collisions are rare and, we hope, may be neglected. The above discussion of the cut-off procedure is heuristic and much effort has been expended on arguments which are mathematically more respectable, with a good deal of success; Ecker[6] gives great detail.

To proceed with the result, we have, by definition,

$$b_\perp = \frac{|ee_s|}{4\pi\epsilon_0 m u^2} \qquad 8.32$$

Then, for the minimum value of θ, symbol θ_m, we find

$$\lambda_D = b_\perp \cot(\theta_m/2) \qquad 8.33$$

$$\text{or} \quad \sin(\theta_m/2) = b_\perp/\sqrt{(b_\perp^2 + \lambda_D^2)} \qquad 8.34$$

Using equation 8.34 in equation 8.31, the result is $\ln[1/\sin(\theta_m/2)]$

$$\text{or} \quad \ln \sqrt{\frac{b_\perp^2 + \lambda_D^2}{b_\perp^2}} \approx \ln\left(\frac{\lambda_D}{b_\perp}\right) = L \qquad 8.35$$

Equation 8.35 defines what is called the 'Coulomb logarithm'. For normal plasmas, it has numerical values in the range 10–20 and it has the advantage of being only slightly sensitive to the actual values of λ_D and $b_\perp$. For example, $b_\perp$ is a function of u, and this lack of sensitivity means that we can take average values, etc.

With a method of handling the evaluation of the Coulomb integral, we are in a position to make realistic calcualtions of the energy and momentum losses due to collisions. We suppose that there are several species of field particles α whose distribution functions are denoted by $f_\alpha(\mathbf{v}_s)\, d\mathbf{v}_s$. It will frequently be convenient to normalize f_α to unity. Putting in the constant R^2, we can then write

$$\left\langle\frac{\mathrm{d}p}{\mathrm{d}t}\right\rangle = \frac{-e^2}{4\pi\epsilon_0{}^2}\sum_\alpha \frac{e_s{}^2}{\mu U^3}\,U\,.\,L\,f_\alpha\,(\mathrm{v}_s)\mathrm{d}\mathrm{v}_s \tag{8.36}$$

$$\left\langle\frac{\mathrm{d}E}{\mathrm{d}t}\right\rangle = \frac{-e^2}{4\pi\epsilon_0{}^2}\sum_\alpha \frac{e_s{}^2}{\mu U^3}\,C\,.\,U\,.\,L\,f_\alpha(\mathrm{v}_s)\,\mathrm{d}\mathrm{v}_s \tag{8.37}$$

Rosenbluth, MacDonald and Judd[9] pointed out an elegant way of solving these expressions. Consider, for example, the integral for one species as it occurs in equation 8.36. We have to evaluate

$$\iiint f(\mathrm{v}_s)\,\frac{(\mathrm{v}-\mathrm{v}_s)}{|\mathrm{v}-\mathrm{v}_s|^3}\,.\,\mathrm{d}v_{1s}\,\mathrm{d}v_{2s}\,\mathrm{d}v_{3s}$$

which is exactly the same form as the expression

$$E(r) = \frac{1}{4\pi\epsilon_0}\int \rho(r_s)\,.\,\frac{r-r_s}{|r-r_s|^3}\,\mathrm{d}r$$

which gives the electrostatic field due to a charge distribution $\rho(r_s)$. The difference is that in the plasma case we are dealing with $f_\alpha(v_s)$, a distribution defined in velocity space, while in electrostatics we deal with a distribution in ordinary co-ordinate or laboratory space.

Since many field problems in electrostatics have well-known solutions, the analogy can be used to save time, as well as having an inherent physical basis. Some points from electrostatics will shortly emerge, but for the moment let us continue with the details. To use the analogy, let

$$\rho_\alpha = \frac{(ee_s)^2}{4\pi\epsilon_0{}^2}\cdot f_\alpha\,(\mathrm{v}_s) \tag{8.38}$$

Then, the 'potential' defined in velocity space is

$$\phi_\mathrm{v} = \int \frac{\rho_\alpha}{|\mathrm{v}-\mathrm{v}_s|}\,\mathrm{d}\mathrm{v}_s \tag{8.39}$$

and $$E_\mathbf{v} = -\nabla_\mathbf{v}\,.\,\phi(v) = \int \frac{U}{|U|^3}\,\rho_\alpha\,\mathrm{d}\mathrm{v}_s \tag{8.40}$$

Using these,

$$\left\langle\frac{\mathbf{d}p}{\mathrm{d}t}\right\rangle = -\sum_\alpha \frac{L}{\mu}E_\mathbf{v} \tag{8.41}$$

$$\left\langle\frac{\mathrm{d}E}{\mathrm{d}t}\right\rangle = -\sum_\alpha L\left\{\frac{\mathrm{v}\,.\,E}{\mu}-\frac{\phi}{m}\right\} \tag{8.42}$$

where we eliminated C in equation 8.42.

For completeness, let us derive the expression for the rate of change of the mean square perpendicular momentum, the mean perpendicular momentum obviously being zero. We have

$$p^2 = p_\perp^2 + p_=^2$$

and $$\frac{d(p^2)}{dt} = 2m\,\frac{dE}{dt}$$

$$\frac{d(p_=^2)}{dt} = 2p_=\frac{dp_=}{dt} = 2p\,\frac{dp}{dt}$$

$$\therefore \quad \left\langle\frac{d(p_\perp^2)}{dt}\right\rangle = 2m\left\langle\frac{dE}{dt}\right\rangle - 2p\,\frac{dp}{dt}$$

$$= 2\sum_{\alpha} L\,.\,\phi_V \qquad 8.43$$

on using equations 8.41 and 8.42. This shows a gain of transverse momentum while losing forward momentum.

The equations for ϕ_V and E_V are soluble in many important cases, so that we are now able to calculate the relaxation times. For example, the relaxation time for the parallel momentum is given by

$$\tau_= \frac{-p_=}{\langle dp_=/dt\rangle} = \frac{mv}{\langle dp_=/dt\rangle} \qquad 8.44$$

The energy relaxation time and the relaxation time of $p_\perp^2$ are defined in an exactly similar manner.

Example

A fast electron is fired into a field of test particles, which are electrons characterized by an isotropic Maxwellian velocity distribution. Find the slowing-down time, i.e. the time in which the test electron is brought to rest.

Here, we must evaluate $\langle dp_=/dt\rangle$ for use in equation 8.44.

$$f(\nu_s) = N_e\left(\frac{m}{2\pi kT_e}\right)^{3/2}\exp\left(-\frac{mv_s^{\,2}}{2kT_e}\right)$$

which we abbreviate to $N_e(\beta^3/\pi^{3/2})\exp(-\beta^2\,\nu_s^{\,2})$.

This distribution is obviously symmetrical in velocity space so, according to Gauss's theorem, the electric field at $r = \mathrm{v}$, is that due to the charge *inside* the sphere radius r. Physically this means that only the *slow* field electrons scatter the fast test electrons. The expression for E_V is

$$E_V = \frac{v}{v^3}\frac{(ee_s)^2}{4\pi\epsilon_0{}^2} \cdot N_e \int_0^v 4\pi r^2 f^1(r)\, dr^*$$

$$= N_e \frac{(ee_s)^2}{4\pi\epsilon_0{}^2} \cdot \frac{v}{v^3} \cdot \Phi_1(\beta v) \qquad 8.45$$

where $\Phi_1(z) = \Phi(z) - z\,\Phi^1(z)$

$$\Phi(z) = \text{erf}\, z = \frac{2}{\pi^{1/2}} \int_0^z \exp(-\zeta^2)\, d\zeta \qquad 8.46$$

Let us also evaluate ϕ_V, although we do not immediately require it. Again, using the electrostatic analogy,

$$\phi_V = \frac{1}{v}\int_0^v \rho_\alpha\,(dv_s) + \int_v^\infty \frac{1}{v_s}\rho_\alpha\, d(v_s)$$

giving $\phi_V = \frac{(ee_s)^2}{4\pi\epsilon_0{}^2} \cdot \frac{N_e}{v} \cdot \Phi(z)$ 8.47

It will be useful to remember that both $\Phi(z)$ and $\Phi_1(z)$ tend to unity with large z. Since, for electrons, $\mu = m/2$ and $(ee_s)^2 = e^2$, we can write down the final result that

$$\tau_= = \tfrac{1}{2}\left(\frac{4\pi\epsilon_0{}^2}{e^4}\right)\frac{m^2 v^3}{LN_e \Phi_1(\beta v)} \qquad 8.48$$

Exercises

1. Show that $\tau_\perp$ is given by

$$\tau_\perp = \frac{1}{4}\left(\frac{4\pi\epsilon_0{}^2}{e^4}\right) \cdot \frac{m^2 v^3}{LN_e \Phi(\beta v)} \qquad 8.49$$

It will be noticed that $\tau_= \rightarrow 2\tau_\perp$, for large βv.

2. Find the expression for the energy relaxation time. Some care is needed here.

The results for ions are similar, except that factors of $(M_i/m)^{1/2}$ appear in the momentum relaxation times and one of (M_i/m) in the energy relaxation. This means that a disturbance relaxes first by establishing an equilibrium Maxwellian distribution for the electrons, at some temperature T_e. Next, the ions reach equilibrium at T_i after a considerably longer time and, last, after a still longer time, T_e and T_i reach a

*Integrals like this are easily done by observing that $\int r^2 \exp(-\beta^2 r^2)\, dr = \frac{d}{d(\beta^2)} [\int \exp(-\beta^2 r^2)\, dr]$
$z\,\Phi^1(z)$ also equals $(2z/\sqrt{\pi}) \exp(-\beta^2 r^2)$.

common equilibrium. The discussion of special cases and numerical values is the province of plasma physics and many examples are given in the references.

Chandrasekhar[10] is very useful in discussing the behaviour of the functions involved when it is improper to approximate using the asymptotic values. It is not superfluous to remark that he was led to expressions of the type discussed here by consideration of the behaviour of stars in the galaxy, where the gravitational force plays the role of the Coulomb force. Luckily, the time scales are rather different! The articles by Trubnikov[11] and Sivukhin[8] give many applications of relaxation-time theory in plasma physics. Particularly important phenomena in reactor physics, such as electron runaway and particle loss, can be analysed rather directly, i.e. without the actual solution of a kinetic equation, if enough is known about the relaxation times.

The potentials, as defined above, are in the form used by Trubnikov and by Sivukhin. In the next section we discuss a new kinetic equation or, rather, a new form of the collision integral, which implies a slightly different form for these quantities, being that originally prescribed by Rosenbluth et al.[9] The derivation is informative at several levels.

8.6 The Fokker–Planck form for the collision integral

In section 6.7, we gave a general physical description of the ideas underlying the Fokker–Planck equation, the most important of which is that it describes the motion of a particle which undergoes large numbers of weak collisions of such a nature that the particle 'remembers' only the penultimate collision and not earlier ones. In equation 6.52 we introduced Smoluchovski's equation, which we now write with the velocity $\mathbf{v}$ instead of the state λ used earlier. If the velocity changes from $\mathbf{v}-\Delta\mathbf{v}$ to $\mathbf{v}$ in the time Δt, the equation reads

$$f(\mathbf{v}, t+\Delta t) = \int f(\mathbf{v}-\Delta\mathbf{v}) \,.\, W(\mathbf{v}-\Delta\mathbf{v}, \nabla\mathbf{v})\, \mathrm{d}\mathbf{v} \qquad 8.50$$

where it will be remembered that W is the probability of the transition $\mathbf{v}-\Delta\mathbf{v}\rightarrow\mathbf{v}$ occurring in Δt.

The integral in equation 8.50 can be expanded to

$$f(\mathbf{v}\,.\,t)\, W(\mathbf{v}, \Delta\mathbf{v})_{\Delta t} - \Delta\mathbf{v}\,.\,\frac{\partial}{\partial\mathbf{v}}\,(fW_{\Delta t}) + \tfrac{1}{2}\,(\Delta\mathbf{v}\,.\,\partial/\partial\mathbf{v})^2\,(fW_{\Delta t})\ldots\text{etc.} \qquad 8.51$$

Integrating equation 8.51 means taking the averages over W, so we get

$$\frac{f(\mathbf{v}, t+\Delta t) - f(\mathbf{v}, t)}{\Delta t} = -\frac{\partial}{\partial\mathbf{v}}\,.\,[f\langle\Delta\mathbf{v}] + \tfrac{1}{2}\frac{\partial^2}{\partial\mathbf{v}\,.\,\partial\mathbf{v}}\,(f\langle\Delta\mathbf{v}\,.\,\Delta\mathbf{v}\rangle)$$

and, in the limit $\Delta t\rightarrow 0$,

$$\left(\frac{\partial f}{\partial t}\right)_{\mathrm{c}} = -\frac{\partial}{\partial\mathbf{v}}\,[f\langle\Delta\mathbf{v}\rangle] + \tfrac{1}{2}\,\frac{\partial^2}{\partial\mathbf{v}\,.\,\partial\mathbf{v}}\,(f\langle\Delta\mathbf{v}\,.\,\Delta\mathbf{v}\rangle) \qquad 8.52$$

In equation 8.52,

$$\langle \Delta \mathbf{v} \rangle = \frac{1}{\Delta t} \int W(\mathbf{v}, \Delta \mathbf{v})\, \Delta \mathbf{v}\, \mathrm{d}\Delta \mathbf{v} \qquad 8.53$$

$$\langle \Delta \mathbf{v} \,.\, \Delta \mathbf{v} \rangle = \frac{1}{\Delta t} \int W(\mathbf{v}, \Delta \mathbf{v})\, [\Delta \mathbf{v} \,.\, \Delta \mathbf{v}]\, \mathrm{d}\Delta \mathbf{v} \qquad 8.54$$

Here, equation 8.52 is called the 'Fokker–Planck collision term'. We shall shortly see that equation 8.53 defines the coefficient of dynamical friction and equation 8.54 the diffusion coefficient, named by analogy with the similar quantities in Chapter 6.

Equation 8.52 gives an expression for the right-hand term of the Boltzmann equation, which ought to be valid when large numbers of weak collisions take place. But this is also the situation which we studied in the last section when we calculated the relaxation times, and so the two approaches cover the same range of plasma conditions. Because the theoretical foundation of equation 8.52 is better than that of the transport equation with relaxation term, it is very satisfactory that it turns out that there are very close relationships between the dynamical friction and diffusion coefficients and the relaxation times, so that the two approaches lead to the same conclusions. Note that the Fokker–Planck collision term is non-linear and that solution of the full transport equation using this form is not usually possible without extensive computation.

The value of the treatment is mainly bound up with the actual form of the r.h.s., which contains much useful information. Let us now discuss this aspect of the equation, remembering that we are discussing the behaviour of particles in momentum or velocity space rather than in laboratory space. When we introduced the Fokker–Planck equation in Chapter 6, it was mentioned that it was physically a generalized continuity equation. Here we can write the l.h.s. of the Boltzmann equation, taken in velocity space and neglecting collisions, as

$$\frac{\partial f_\alpha}{\partial t} + \mathrm{div}_{\mathrm{r}}\,(\mathbf{v} f_\alpha) + \mathrm{div}_{\mathrm{v}}\,(j) = 0 \qquad 8.55$$

where $j = (F_\alpha/m_\alpha) f_\alpha = \dot{\mathbf{v}} f_\alpha$ is the current (in velocity space) due to the external force F. Now we consider collisions, and in particular the manner in which they modify the flux of j. If we consider a small box of particles in the velocity space, the motion due to the external force will set up a flow through the box, represented by the third term of equation 8.55. If collisions also occur, some particles near to the box will jump into it (and particles inside the box will jump out), which is merely how diffusion works. The number of such particles is proportional to the number density, evaluated for example at the centroid of the (small) box, and to a diffusion coefficient which turns out to be a tensor quantity. Thus, it is proper to write

$$j = \dot{\mathbf{v}} f_\alpha - D_{\mathrm{ik}} \cdot \nabla_{\mathrm{v}} f_\alpha \qquad 8.56$$

which is analogous to the expression usual for particles in coordinate (laboratory) space:

$$j = \rho \mathbf{v} - D \nabla_{\mathbf{v}} \rho \tag{8.57}$$

In equation 8.36 $\dot{\mathbf{v}} f_\alpha$ has two components where collisions are included, one of which is due to the externally applied force and the other being an internal force which represents the frictional force due to collisions. Making this division, equation 8.55 can be put in the form

$$\frac{\partial f_\alpha}{\partial t} + \nabla_{\mathbf{r}} (\mathbf{v} f_\alpha) + \nabla_{\mathbf{v}} \left(\frac{F_{\text{ext.}}}{m_\alpha} f_\alpha \right) = - \nabla_{\mathbf{v}} j_\alpha \tag{8.58}$$

$$\text{where} \quad j_\alpha = \frac{F_{\text{int.}}}{m_\alpha} f_\alpha - D_{\text{ik}} \frac{\partial f_\alpha}{\partial \mathbf{v}} \tag{8.59}$$

and where the r.h.s. of equation 8.58 is purely collisional and F_{int} in equation 8.59 is the frictional force. Equation 8.59 is, of course, essentially the same as equation 8.52 but it renders the physics of the Fokker–Planck equation much more obvious and elucidates the appropriate nature of the names given to the coefficients.

Next, consider the evaluation of the coefficients. First, we shall reduce the r.h.s. of equation 8.52 to a form which is very frequently encountered in the literature and which uses the velocity-space potentials in the manner introduced by Rosenbluth et al.[9] One of these is nearly identical with $\phi_{\mathbf{v}}$ defined by equation 8.39; the other is new. The usual symbols and definitions are

$$H(\mathbf{v}) = \left(1 + \frac{m}{m_{\text{s}}}\right) \int \frac{\mathrm{d}\mathbf{v}_{\text{s}} f(\mathbf{v}_{\text{s}})}{|\mathbf{v} - \mathbf{v}_{\text{s}}|} \tag{8.60}$$

or, for scattering by the same mass,

$$H(\mathbf{v}) = 2 \int \frac{\mathrm{d}\mathbf{v}_{\text{s}} f(\mathbf{v}_{\text{s}})}{|\mathbf{v} - \mathbf{v}_{\text{s}}|} \tag{8.61}$$

The new one is

$$G(\mathbf{v}) = \int \mathrm{d}\mathbf{v}_{\text{s}} \,.\, f(\mathbf{v}_{\text{s}}) \, |\mathbf{v} - \mathbf{v}_{\text{s}}| \tag{8.62}$$

Remembering that in equation 8.50 we are discussing changes of magnitude $\Delta \mathbf{v}$ in time Δt, the quantity we wish to evaluate is $\langle \Delta \mathbf{v} / \Delta t \rangle$ in the notation of the last section. This is merely $(1/m) \langle \Delta \boldsymbol{p} / \Delta t \rangle$, so we can rewrite equation 8.36 as follows:

$$\langle \Delta \mathbf{v} \rangle = -K \left(\frac{e_{\text{s}}}{e} \right)^2 \left(1 + \frac{m}{m_{\text{s}}} \right) \int \frac{(\mathbf{v} - \mathbf{v}_{\text{s}})}{|\mathbf{v} - \mathbf{v}_{\text{s}}|^3} f(v_{\text{s}}) \, \mathrm{d}v_{\text{s}} \tag{8.63}$$

$$K = \frac{e^4 L}{4 \pi \epsilon_0{}^2 m^2} \tag{8.64}$$

Using equation 8.60,

$$\langle \Delta \mathbf{v} \rangle = -K \left(\frac{e_s}{e}\right)^2 \frac{\partial H(\mathbf{v})}{\partial \mathbf{v}} \qquad 8.65$$

The evaluation of $\langle \Delta v.\ \Delta v \rangle$ in terms of $G(v)$ entails several manipulations and is relegated to Appendix 8. The result is

$$\langle \Delta \mathbf{v}\ .\ \Delta \mathbf{v} \rangle = K\left(\frac{e_s}{e}\right)^2 \frac{\partial^2 G(\mathbf{v})}{\partial \mathbf{v} \partial \mathbf{v}} \qquad 8.66$$

The Fokker–Planck collision term now reads

$$\left(\frac{\partial f}{\partial t}\right)_c = K\left[-\left(\frac{e_s}{e}\right)^2 \frac{\partial}{\partial \mathbf{v}}\left\{f(\mathbf{v})\frac{\partial H(\mathbf{v})}{\partial \mathbf{v}}\right\} + \frac{{e_s}^2}{2e^2}\frac{\partial^2}{\partial \mathbf{v}\partial \mathbf{v}}\left\{f(\mathbf{v})\ .\ \frac{\partial^2 G(v)}{\partial \mathbf{v}\partial \mathbf{v}}\right\}\right] \qquad 8.67$$

If one now writes $f(\mathbf{v}) = \delta(\mathbf{v} - C)$, i.e. a function which vanishes everywhere except at $\mathbf{v} = C$, one can then take moments of equation 8.67 by multiplying by the desired quantity and integrating in velocity. Then, H and G can be determined for the distribution function $f(\mathbf{v}_s)$. As an example, consider the determination of $\tau_=$ already found in equation 8.48. The steps are, multiply equation 8.67 by $\mathbf{v}$, insert the delta function and integrate over velocity space. This results in

$$\frac{1}{K}\frac{\partial C}{\partial t} = \frac{\partial H(C)}{\partial C} = \frac{C}{C} \cdot \frac{\partial H}{\partial C} \qquad 8.68$$

$$\text{and} \quad \tau_= = -\frac{C}{\partial C/\partial t} = -\frac{C^2}{KC\ .\ \partial H/\partial C} \qquad 8.69$$

For the Maxwellian distribution used to derive equation 8.45,

$$H(\mathbf{v}) = \left(1 + \frac{m}{m_s}\right)\frac{N_e}{v}\Phi(z) \qquad 8.70$$

$$G(\mathbf{v}) = N_e\left[\left(v + \frac{1}{2\beta^2 v}\right)\Phi(z)\ \exp\frac{(-\beta^2 v^2)}{\beta\sqrt{\pi}}\right] \qquad 8.71$$

Differentiation of equation 8.70 and substitution in equation 8.69 leads back to equation 8.48.

For ease of reference we quote the values of

$$\frac{\partial H}{\partial C} = -2N_e\left(1 + \frac{m}{m_s}\right)\beta^2\frac{\Phi_1(z)}{2z^2} \qquad 8.72$$

$$\frac{\partial G}{\partial C} = N_e\left(\Phi(z) - \frac{\Phi_1(z)}{2z^2}\right) \qquad 8.73$$

Here, $\Phi_1(z)$ is the function defined in equation 8.46.

By taking the moment of equation 8.67 for the square of the transverse velocity component, we find that $\tau_\perp$ is given by an expression which is not quite the same as equation 8.49. $\Phi(\beta v)$ in equation 8.49 is replaced by $\Phi(\beta v) - \Phi_1(\beta v)/\beta^2 v^2$, which means the expressions are the same when β is very large, an assumption built into equation 8.49.

Lastly, the energy relaxation time becomes

$$\tau_E = \frac{C^3}{8KN_e\,[\Phi_1'(z)/2z^2]} \tag{8.74}$$

Some further properties of the Fokker–Planck formulation are briefly discussed in Appendix 8.

8.7 Some further kinetic theories

As early as 1936, Landau[13] derived a kinetic equation which occurs frequently in the literature. In our notation, the collision term is

$$\left(\frac{\partial f}{\partial t}\right)_c = \frac{K}{2}\sum_\alpha \left(\frac{e_\alpha}{e}\right)^2 m\frac{\partial}{\partial \mathbf{v}}\int \frac{\partial^2 |\mathbf{v}-\mathbf{v}_s|}{\partial \mathbf{v}\,.\,\partial \mathbf{v}}\left[\frac{f_1(\mathbf{v}_s)}{m}\frac{\partial f_1(\mathbf{v})}{\partial \mathbf{v}} - \frac{f_1(\mathbf{v})}{m}\frac{\partial f_1(\mathbf{v}_s)}{\partial \mathbf{v}_s}\right] \tag{8.75}$$

where $f_1(\mathbf{v}_s) = f(r, \mathbf{v}_s, t)$, $\quad f_1(\mathbf{v}) = f(\mathbf{v}, \mathbf{v}, t)$.

This equation can be shown to be identical with the Fokker–Planck collision term. Many transport calculations based on this equation are given in the literature, so it is valuable to remember that they should be equivalent to similar quantities calculated using the other formulation. An extensive review of results is given by Braginskii[14].

A later and more sophisticated kinetic equation is called the Lenard–Balescu equation[15, 16]. This equation, very roughly, may be said to take the physical problem of computing the range of the Coulomb forces into account in a manner which differs from the, admittedly rather approximate, reasoning based on the Debye length. Instead, the effective potential due to a charge allows for a change in permittivity from ϵ_0 (the vacuum value) to a value which is more nearly that of the plasma. Expressions for such a permittivity were introduced when we discussed the Vlasov equation.

Finally, the reader ought to know that one of the most fruitful techniques in recent kinetic theory has been the formulation of a hierarchical structure of kinetic equations, which is theoretically exact if an infinite sequence of equations is solved. Naturally, it is necessary to cut off the sequence, but even so the method shows that the (valid) kinetic equations are members of the hierarchy. The method is called the Born–Bogliubov–Kirkwood–Green–Yvon method – BBKGY for short! Bogliubov[17] is the standard derivation. The reader will find these matters fully discussed by Balescu[18], Liboff[19] and Wu[20].

Conclusion

While recent years have seen major advances in plasma kinetic theory, it must be said that, although transport calculations are much improved, little progress in the actual solution of the kinetic equation has been made. Another restriction ought to be mentioned. In the fully ionized gas *theory*, there are no ionizing collisions and no emission of radiation; yet the first thing one notices on seeing a laboratory plasma is the strong radiation. A realistic plasma kinetics must, therefore, include radiation. Progress in this field is based on the consideration of a fluctuating electromagnetic field which is coupled to the plasma. This theory is beyond the scope of this book.

9

Electrical and Optical Noise

From the technical point of view, by far the most important instances of statistical fluctuation are first the stochastic signals, called 'noise', encountered in electric circuits and secondly, as techniques for communication are pushed into the optical region, the noise encountered when optical signals are converted into electrical signals, as in present-day systems they must be before becoming useful. These problems are sufficiently important for a detailed discussion to be required.

First, we require a simple method by which we can distinguish between 'noise' and wanted signals. This is easily done by remembering the concept of a stationary random process: one in which the energy in a long time interval T is independent of the time at which T begins. Speech and music, which in non-technical terminology often constitute noise or at least unwanted disturbance, obviously do *not* have this property and therefore are distinguishable from technical noise. More seriously, disturbances such as atmospheric discharges, intermittent crackling due to poor contacts, etc. are not noise and have to be treated differently.

A second important factor is that random noise in electrical systems can result from a number of different causes. The most important, by far, are the fluctuations associated with the dissipative loss in the input circuits of the apparatus and the fluctuations in the number of charge carriers flowing in the input active devices. However, noise can be caused by different phenomena, which may be dominant in particular cases. Examples are the so-called 'flicker' or $1/f$ noise experienced in very-low-frequency amplifier systems, which is even now not fully understood, and noise originating in an input transducer. In fact, the noise behaviour of optical systems is in this class because, in the detection of optical signals, one usually goes from the optical signal to a much-lower-frequency electrical signal at some stage of the transmission system, and both the statistics of the incoming photons and the statistics of the resultant electron flow influence the observed noise.

In this chapter we shall be concerned mainly with electrical noise from resistors and active devices and with the optical detector, and will have little to say on other types of noise. Our aim will be, as far as possible, to reduce the calculation of noise behaviour to a few basic principles of stochastic theory, mainly correlation in time

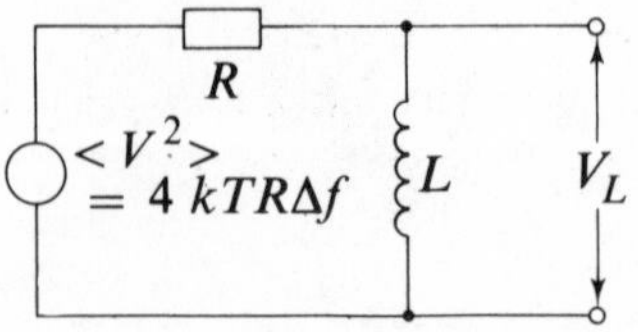

Fig. 9.1 Circuit for technical calculation of noise voltage across L.

and the Wiener–Khinchin theorem and the very significant fluctuation-dissipation theorem.

9.1 The technical method of solving noise problems

To introduce the subject, let us solve a very simple noise problem by the technique used by circuit engineers. A study of the steps involved will clarify the real problems. The circuit of Fig. 9.1 shows a series R–L circuit. The generator is associated with the total loss resistance of the circuit, R. The output is the voltage measured across L.

The steps are as follows.

1. Associate a mean-square voltage with the resistor. According to Nyquist's theorem, the magnitude is $\langle V_n{}^2\rangle = 4kTR\ \Delta f$ measured in Δf at *any f*.

2. Solve the circuit equation using impedance to obtain

$$V_L = \frac{j\omega L\,\langle V_n{}^2\rangle^{\frac{1}{2}}}{Z}.$$

or
$$\langle V_L{}^2\rangle = \frac{\omega^2 L^2}{R^2+\omega^2L^2}\,.\,4kTR\ \Delta f \qquad 9.1$$

$$= \frac{Q^2}{1+Q^2}\,.\,4kTR\ \Delta f \qquad 9.2$$

if we put $Q = \omega L/R$; i.e. $\langle V_L{}^2\rangle$ is slightly less than $\langle V_n^2\rangle$ except at low frequencies.

We notice that the response, equation 9.1, can be written as $|G(\omega)|^2\,.\,4kTR\ \Delta f$ or $|G(\omega)|^2\,.\langle V_n^2\rangle$ and therefore the inductor only enters into the expression for $G(\omega)$.

Step 2 is readily disposed of: the impedance technique is simply a shorthand for solving the circuit differential equation

$$V(t) = i(t)\,R + L\,\frac{di(t)}{dt} \qquad 9.3$$

by Fourier transformation and requires no discussion here. Step 1, however, requires more discussion.

Johnson empirically established that a purely metallic resistor, resistance R, was the source of a mean square noise voltage $\langle V_n{}^2\rangle = 4kTR\ \Delta f$, and Nyquist was able to justify the result theoretically by arguments based on classical equipartition theory applied to a known length of properly terminated transmission line in a constant-temperature oven, but the connection of this with physical statistics seems a bit remote. Moreover, it is far from clear why the noise generator is associated only with the resistor and not with the inductor (except in so far as a real inductor has a loss resistance which, in our calculation, was taken into R). We begin to see the answer if we rewrite equation 9.3 in the standard form

$$\frac{di}{dt} + \frac{R_i}{L} = V(t)\,/\,L$$

which is exactly that of the Langevin equation for Brownian motion, equation 6.19, with R/L equivalent to β/m. Thus, this coefficient determines the magnitude of the fluctuations, as it did in the Brownian case. The only way the damping coefficient can be made negligible for all finite frequencies is by letting $R \to 0$. This is an argument for a very intimate connection between resistance and fluctuation amplitude. Furthermore, any relationship we find must be consistent with the equipartition theorem $\frac{1}{2}L\langle i^2\rangle = \frac{1}{2}kT$, at least for low frequencies where there can be no quantum effects. As a preliminary to deriving the above results directly from the Langevin equation, it is worthwhile seeing how Nyquist exploited equipartition theory in his original derivation of his theorem.

9.2 Nyquist's theorem: classical version

This derivation refers to the system shown in Fig. 9.2. A length of lossless transmission line of characteristic impedance $Z_0 = \sqrt{(L/C)}$ is terminated at both ends

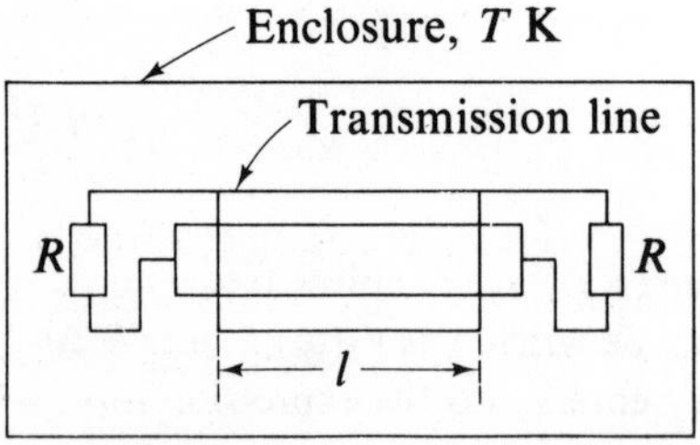

Fig. 9.2 Arrangement used by Nyquist to derive his theorem.

with metallic resistors R, adjusted so that $R = Z_0$. There is then no reflection at the terminations and no standing waves on the line. This system is in a heat bath at temperature T and has been there sufficiently long for thermal equilibrium to be established.

Nyquist[1] considered *only* the *TEM* electromagnetic modes, counted their number in a given frequency interval and then associated energy of $\frac{1}{2}kT$ with the E and H fields of each mode, to obtain an expression for the power in the line. Since the system is in equilibrium, half this power is associated with each resistor and this leads to the expression $\langle V_n^2\rangle = 4kTR\ \Delta f$. The details are as follows.

The fundamental *TEM* resonant frequency of the line is

$$f_1 = c/2l \tag{9.4}$$

where c = phase velocity on the line = velocity of light for lossless line.

In general, the resonances are given by

$$f_n = nc/2l = nf_1, \quad n = 1, 2, 3, 4, \text{etc.}$$

Thus, in a frequency interval Δf at f_1 there are ΔN resonances where

$$\Delta N = \frac{\Delta f}{f_1} = \frac{2l}{c}\,\Delta f \qquad 9.5$$

Each *resonance* makes a contribution of kT to the energy; therefore the energy in the range Δf is equal to $(2l/c)\,kT\,\Delta f$. But, in the line considered, the group velocity equals the phase velocity, owing to the assumption of zero loss. Therefore, a wave takes l/c seconds to travel from one end of the line to the other and the power in the line is simply

$$P = 2kT\,\Delta f \text{ watts} \qquad 9.6$$

Numerically, $kT\,\Delta f = 4.0 \times 10^{-21}$ watts/cycle at ambient temperature and each resistor is responsible for half this power. Since, viewed from one resistor, the input impedance of the line is R, we can immediately write

$$\langle V_n^2 \rangle = 4kTR\,\Delta f \qquad 9.7$$

The argument has a number of unsatisfactory features. Firstly, the line can also resonate in *TE* and *TM* modes and it is not obvious that the counting done above is really correct. This objection can be removed by considering a waveguide which is incapable of supporting *TEM* modes, and then counting the *TE* and *TM* modes: the same result is obtained. Secondly, the argument is classical and does *not* allow for quantum effects. This we shall overcome later by an argument using Planck's law and it will be shown that equation 9.7 is obtained when $\hbar\omega \ll kT$. The most serious objection is that the calculation makes no real identification of the source of the noise. Basically, it assumes that the fluctuations are linked to the dissipation.

9.3 Solution using correlation functions

To investigate this last point more deeply, we shall solve the Langevin equation using the technique of correlation functions. We write the equation, still for the circuit of Fig. 9.1, in the form

$$\frac{di}{dt} + \frac{i}{\tau_c} = V(t)/L, \quad \tau_c = \frac{L}{R} \qquad 9.8$$

with the understanding that $V(t)$ is shorthand for a stochastic process in which[2]

$$\langle V(t) \rangle = 0, \quad \langle V(t)^2 \rangle \neq 0$$

Integrating 9.8 in time, we get

$$i = i_0 \exp\left(-\frac{t}{\tau_c}\right) + \exp\left(-\frac{t}{\tau_c}\right) \int_0^t \exp\left(\frac{t'}{\tau_c}\right) V(t')/L \, dt'$$

We now form the current-correlation function:

$$\langle i(t)\,.\,i(t+\tau)\rangle = \left\langle \left(i_0 \exp\left(-\frac{t}{\tau_C}\right) + \exp\left(-\frac{t}{\tau_C}\right) \int_0^t \exp\left(\frac{t'}{\tau_C}\right) V(t')/L\,\mathrm{d}t' \right) \times \right.$$

$$\left. \left(i_0 \exp\left[-\frac{(t+\tau)}{\tau_C}\right] + \exp\left[-\frac{(t+\tau)}{\tau_C}\right] \int_0^{t+\tau} \exp\left(\frac{t''}{\tau_C}\right) V(t'')/L\,\mathrm{d}t'' \right) \right\rangle$$

$$= i_0^2 \exp\left[-\left(\frac{2t+\tau}{\tau_C}\right)\right] + \exp\left[-\left(\frac{2t+\tau}{\tau_C}\right)\right] \int_0^t \mathrm{d}t' \int_0^{t+\tau} \mathrm{d}t'' \exp\left(\frac{t'+t''}{\tau_C}\right) \langle V(t')\,.\,V(t'')\rangle/L^2 \qquad 9.9$$

because of the condition on $\langle V(t)\rangle$. The term $\langle V(t')\,.\,V(t'')\rangle$ is the autocorrelation function of the voltage driving the circuit. We remember that it is a function only of $t''-t'=s$ and that it is symmetrical about $s=0$. Write it as $K_V(s)$. Put $t'+t''=u$; then the repeated integral* becomes

$$\tfrac{1}{2} \int_\tau^{2t+\tau} \exp(u/\tau_C)\,\mathrm{d}u \int_{-\tau}^{+\tau} K_V(s)\,\mathrm{d}s$$

But $K_V(s)$ is very sharply peaked near $s=0$, therefore the limits can be extended to $\pm\infty$ in the second integral. Equation 9.9 is thus equal to

$$\langle i(t)\,.\,i(t+\tau)\rangle = i_0^2 \exp\left[-\frac{(2t+\tau)}{\tau_C}\right] + \frac{\tau_C}{2L^2}\left(1-\exp\left[-\frac{2t}{\tau_C}\right]\right) \int_{-\infty}^{+\infty} K_V(s)\,\mathrm{d}s \qquad 9.10$$

If we now let $t \to 0$, we find

$$\langle i(o)\,.\,i(\tau)\rangle = i_0^2 \exp\left(-\frac{|\tau|}{\tau_C}\right) = i_0^2 \exp\left(-\frac{R|\tau|}{L}\right) \qquad 9.11$$

for the autocorrelation function of the current in the circuit.

Similarly, when $t \gg \tau$,

$$\langle i^2(t)\rangle = i_0^2 \exp\left(-\frac{2t}{\tau_C}\right) + \frac{\tau_C}{2L^2}\left[1-\exp -\frac{2t}{\tau_C}\right] \int_{-\infty}^{+\infty} K_V(s)\,\mathrm{d}s$$

but t is also much greater than τ_C, so this is equivalent to

$$\langle i^2(t)\rangle = \frac{\tau_C}{2\mathrm{L}^2} \int_{-\infty}^{+\infty} K_V(s)\,\mathrm{d}s \qquad 9.12$$

*Integrals of this type can be evaluated using Fourier or Laplace convolutions, and many useful tables exist.

Equation 9.12 gives the mean-square current when statistical equilibrium has been established in the system. But we have also to satisfy the equipartition theorem which says that $\langle i^2(t)\rangle = kT/L$ in this condition. Thus,

$$R = \frac{1}{2kT}\int_{-\infty}^{+\infty} K_v(s)\, ds \qquad 9.13$$

Equation 9.13 may be looked on as a fluctuation-dissipation theorem for this circuit. For all finite temperatures it shows that the voltage correlation function depends on the value of the resistance in the sense that only for $R = 0$ does $K_v(s) = 0$.

We can go further. When the system has been operating long enough to reach equilibrium, the current autocorrelation function is

$$K_i(\tau) = \langle i(o)\,.\,i(\tau)\rangle = \frac{kT}{L}\exp\left(-\frac{|\tau|}{\tau_c}\right) \qquad 9.14$$

Therefore, using the Wiener–Khinchin theorem,

$$\Phi_i(\omega) = \frac{2kT}{2\pi L}\int_0^{\infty} \exp\left(-\frac{|\tau|}{\tau_c}\right)\exp(-j\omega\tau)\, d\tau$$

or $$\Phi_i(\omega) = \frac{kT}{\pi L}\cdot\frac{\tau_c}{1+\omega^2\tau_c^{\,2}} = \frac{kT}{\pi}\cdot\frac{R}{R^2+\omega^2L^2} \qquad 9.15$$

From the physical meaning of $\Phi_i(\omega)$, if we call $\langle i^2(\omega)\rangle$ the mean-square current fluctuations in the range $\Delta\omega$ at ω we can write

$$\langle i^2(\omega)\rangle = 2\,\Phi_i(\omega)\, d\omega$$

$$= \frac{4kT}{R}\cdot\frac{df}{1+(\omega L/R)^2} \qquad 9.16$$

to which corresponds the voltage fluctuation across the inductor:

$$\langle V^2(\omega)\rangle = 4kTR\, df\,\frac{(\omega L/R)}{1+(\omega L/R)^2} \qquad 9.17$$

which was the result obtained from the technical calculations.

We can consider equations 9.16 and 9.17 in the following way. If $L = 0$, equation 9.11 gives $\langle i^2\rangle = (4kT/R)\, df = 4kTG\, df_1$ where $G = 1/R$ = conductance. If we consider this resistor with an associated current generator whose mean-square amplitude is $4kTG df$, we get the circuit of Fig. 9.3(a). By ordinary circuit transformation, this can be represented as in Fig. 9.3(b) and then, if we wish to connect an inductor in series as Fig. 9.3(c), circuit theory leads to equation 9.12.

We have now, in a particular case, derived Nyquist's result by probabilistic arguments of a very generalized kind. We could repeat the steps for an R–C circuit and, with considerably more mathematical difficulty, for an R–L–C circuit; in each case we would find that we could use the noise voltage or current generator, associating the open-circuit voltage or short-circuit current with a numeric multiplied by the

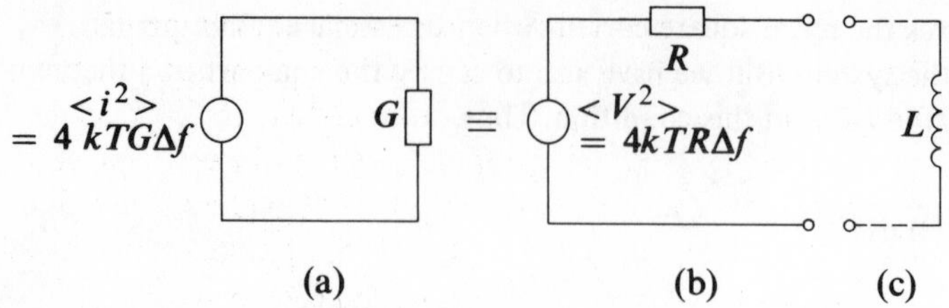

Fig. 9.3 Illustration of current ∿ voltage transformation.

resistance or conductance. It is hardly necessary to say that this fact allows an enormous simplification in the work of solving a real circuit-noise problem and that nobody would *use* the correlation method in simple problems of the type so far outlined.

What of the restrictions? In this section we used the classical equipartition theorem to assign an amplitude to the fluctuation, i.e. $L\langle i_0{}^2\rangle = kT$. As is readily seen, this particular assignment is at fault at very high frequencies, since it assigns to the generator a spectral density which is quite independent of frequency. Therefore, integrating over all frequencies, one could draw infinite power from the resistor, which is nonsense. Quantum effects, dealt with shortly, ensure that the frequency spectrum decays to zero at very high frequencies and the integral is finite.

Some purely mathematical points about the validity of equation 9.5 are discussed in several of the papers in the collection by Wax[2] and need not detain us here. Another point is worth mentioning though. Many writers use an 'ergodic signal theorem' to ensure the identity of time averages and ensemble averages. This seems, to me at least, tautologous since an ergodic signal turns out to be defined as one whose time average *is* equal to the ensemble average – thus stationary random processes are quoted as examples of ergodic signals. However, a careful examination of what one really needs to know in correlation theory shows that all that is required is a knowledge of the *amplitude* of a fluctuation determined from the ensemble which is equated to the amplitude of the fluctuations in time. The evolution of the fluctuations in time is in accordance with appropriate probabilistic *axioms*, for example that one is dealing with a stationary random process*, or other mathematically tractable processes, and there need be no further cross-connection with the ensemble theory. It therefore seems, to me, unnecessary to introduce an ergodic theorem into random signal theory, quite apart from any discussion of its usefulness or lack of it in the theory of molecular chaos. Finally, notice that, just as in any other branch of mathematical physics, the axioms have to be reconsidered if they give rise to theories which disagree with experiment.

9.4 The physical basis of resistor noise

In the last section, we found an expression for the current autocorrelation function, namely $K_i(\tau) = i_0{}^2 \exp(-|\tau|/\tau_c)$, where τ_c is the time constant of the circuit, depen-

*It seems much easier to show experimentally, that one is dealing with a stationary random process than it would be to show the equality of ensemble and time averages.

ding on both the L and R of the circuit. We earlier obtained the same type of expression for the autocorrelation function of the velocity of a Brownian particle, where the time constant depended on the frictional coefficient and the particulate mass. Suppose we consider an isolated resistor and postulate that the equation of motion of the electrons, like that of the Brownian particles, contains a frictional coefficient which is due to collisions, whose nature we need not yet specify. We should then expect the electron current pulses to be correlated according to const. x $\exp(-|\tau|\tau_e)$ and, provided only that τ_e is short enough to make $\omega\tau_e \ll 1$ for all *normal* frequencies, we should still have $\langle i^2(\omega)\rangle = 4kTG\mathrm{d}f$, since equations 9.15 and 9.16 apply with τ_e replacing τ_c. But, in studying the relaxation-time approximation to the Boltzmann theorem, we have already seen that the resistivity/conductivity of metals and semiconductors depends on the density of charges and on the collisional relaxation time. For example, in a simple semiconductor at a fixed temperature, $\sigma = a\tau_e$, say; then equation 9.16 becomes $\langle i^2(\omega)\rangle = 4kTa\tau_e \cdot \dfrac{\mathrm{d}f}{1+\omega^2\tau_e^{\,2}}$ for a block of unit volume. Thus, the current fluctuations can be completely specified in terms of τ_e.

The physical idea behind all this is that the electrons moving between collisions constitute a train of pulses of more or less well-determined shape and duration, which occur at random times. When the resistor is *not* in complete equilibrium but instead a continuous current flows, the mean number of individual pulses per unit time can easily be found. The deviation can be estimated by fluctuation techniques and Fourier analysis applied to the random waveform. Thus, a second technique for noise calculation is suggested.

It is noteworthy that this scheme lends itself readily to the calculation of current noise in active devices. For clarity, we consider a thermionic diode. It is clear that the 'collision' time of importance is the transit time: the time between leaving the cathode and colliding with the anode. In this particular case we have a rather good theoretical estimate of the duration/shape of each pulse, and the calculation can be carried out in considerable detail. In general, however, the same considerations operate in more complicated thermionic devices and in semiconductor devices passing currents.

9.5 Quantum effects in resistor noise

Let us now turn our attention to the problem of modifying Nyquist's formula to take account of quantum effects. In discussing Bose–Einstein statistics, we have already considered that there is an equivalence between the resonant frequencies of a cavity and photons of various energies, and we obtained the following expression for the average energy of a mode of angular frequency ω, real frequency f:

$$\langle E\rangle = hf\left\{\tfrac{1}{2} + \frac{1}{\exp(hf/kT)-1}\right\} \qquad 9.18$$

For Nyquist's lossless transmission line, the group velocity was c and the length l, so the corresponding power is $P = \langle E\rangle c/l$. From equation 9.5 there are two waves, and the polarization conditions $\boldsymbol{E}$ and $\boldsymbol{H}$, in the frequency spacing corresponding with $\Delta N = 1$, i.e.

$$\Delta f = c/2l$$

$$\therefore \quad P = \langle E \rangle \, \Delta f$$

$$= hf\Delta f\left\{\tfrac{1}{2} + \frac{1}{\exp(hf/kT) - 1}\right\} \qquad 9.19$$

If the factor $\frac{1}{2}$ is ignored, this reduces to the earlier value $P = kT\Delta f$ when $hf \ll kT$.

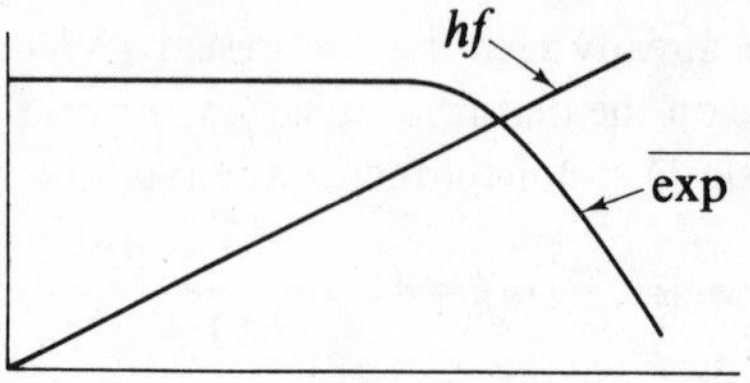

Fig. 9.4 Resistor noise, modified for quantum effects at very high frequencies.

The two terms of equation 9.19 are plotted separately in Fig. 9.4, which shows that at all normal radio frequencies the term $hf/2$ due to the zero-point energy is negligible compared with the second term. However, at sufficiently high (optical) frequencies the situation is reversed. We shall discuss this region in detail later on; here, we merely remark that the zero-point energy cannot be coupled out of the field and is *not* included in ordinary measurements. If we like, we can think of it as due to a generator at 0 K; clearly we cannot extract power from such a generator with a load which is at any finite temperature.

We leave another method of reaching the same result as an example, a fairly difficult one, for the reader. It is relevant to radiometry.

Example

Consider a large black-body radiator and a waveguide cavity receiver, coupled to a horn antenna. Let $G(\theta, \phi)$ be the ratio of the power radiated by the antenna in the (θ, ϕ) direction to the power radiated in the same direction by an isotropic antenna driven by the same power. In symbols,

$$G(\theta, \phi)\, \mathrm{d}\Omega = \frac{P_\omega(\theta, \phi)}{P_\omega} \cdot \frac{\mathrm{d}\Omega}{4\pi}$$

where the P_ω's are radiated powers and $\mathrm{d}\Omega$ the solid angle subtended by the horn.

Let the power received by the horn equal

$$I_\omega(\theta, \phi) \,.\, A(\theta, \phi)\, \mathrm{d}\Omega$$

First, consider the system when the black body and the receiver are in a heat bath and therefore are in thermal equilibrium. Use the result to show that $G(\theta, \phi) = 4\pi\, A(\theta, \phi)/\lambda^2$, where λ = wavelength.

Secondly, show that the maximum available noise power, from the black body, can be written as

$$\frac{c}{4\pi} \cdot \rho(\omega, T)\frac{\lambda^2}{4\pi} \cdot G(\theta, \phi) = \frac{\hbar\omega}{(\exp(\hbar\omega/kT) - 1)} \cdot \frac{G(\theta, \phi)}{2}\,\mathrm{d}f$$

Here, $\rho(\omega, T)$ is the energy density function defined in Chapter 5, and the expression is for *one* direction of polarization.

(The propagation constant in a waveguide, $k = 2\pi/\lambda_g$, is given by $k^2 = \omega^2/c^2 - p_{nm}^2$, where p is a constant for a n, m mode. Hence the group velocity.)

Explain carefully why the radiation emitted by the black body is noise-like.

The above example has introduced the technologically very important concept of the maximum available noise power. It is worth expanding on this idea. In Nyquist's theorem we discussed two resistors in thermal equilibrium and showed that each was responsible for a noise power of $kT\mathrm{d}f$ watts/Hz. If one resistor is held at T K and is connected to an identical resistor at a lower temperature, the hot resistor will transfer power to the cooler one until a new equilibrium is reached. The *maximum* power available for this purpose is $kT\mathrm{d}f$ watts/Hz. We therefore reason that this quantity represents a worst case in design and can be used to estimate the signal–noise performance of amplifiers, etc. If we measure the noise power produced by an amplifier and refer it to the input terminals, obtaining the result $P_n = NkT\mathrm{d}f$ watts/Hz, then the noise figure of the amplifier is defined as N. If the input circuit includes a physical resistor then, very roughly, one unit of this is due to the resistor and $(N-1)$ to the active device.

To conclude this section, let us return to the zero-point energy. We have said that this energy cannot be coupled out of the field: how can it then account for noise in a measuring circuit? The answer is based on Heisenberg's uncertainty principle, which formally states that the measurement of two canonically conjugate variables cannot be better than

$$\Delta p\, \Delta q = \hbar/2$$

Fundamentally, any measurement system tries to measure a p or a q of the incident photon or electromagnetic field pulse. A quantum-mechanical calculation shows that

$$\langle \Delta^2{}_{qf} \rangle = \hbar/4\pi f$$

thus, attempts to measure q will encounter this degree of uncertainty or, in other words, will fluctuate about the mean with this dispersion. The corresponding energy is $\hbar\omega/2$, which is proportional to the power of the fluctuations, so the measurement system appears as though it exhibits a noise source of power $\hbar\omega/2$, even if the driving fields are entirely free from fluctuation. Since this latter requirement involves operation at very high frequencies, the quantum noise energy is correspondingly high.

9.6 Shot noise

In active devices – thermionic or solid-state – the fundamental process is that the current flow through the device is controlled in accordance with signals applied to a control electrode. In general, the currents are due to electrons or holes, so the charge is $\pm|e|$. If the current density is JA/cm^2, J/e charges pass through unit area per second. Current densities cover a very wide range of values, but if we consider $1A/\text{cm}^2$ or $10^4 A/\text{m}^2$, then $0(10^{19})$ charges pass through 1 cm^2 in one second. Thus, if the flow were completely uniform, we should find that exactly one electron would cross the plane in each 10^{-19} second interval. In fact, owing to fluctuations, we observe 0, 1, 2, 3, etc. electrons. Putting the problem another way, if we consider that on average 10^{19} electrons pass the plane in one second, we expect that the measured number will rarely lie outside the range $10^{19}\{1 \pm 1/\sqrt{(10^{19})}\}$. Unfortunately, the fluctuation, although minute by comparison with the mean current, is readily detected by a.c. amplification.

Let us now set up a formal theory for some important cases of the shot effect. The experimental system might be a planar thermionic diode, operating in the temperature-limited regime so that the density of electrons in the cathode–anode space does not influence the probability of emission from the cathode. Then, current densities up to a few A/cm^2 can be drawn with anode potentials of a few tens of volts. The maximum electron velocity might be 6.0×10^6 m/s and the time to cross from cathode to anode might be 10^{-10} s. This time, usually called the 'transit' time, symbol τ, is the fundamental time constant of the process and corresponds with the relaxation or collision times we have encountered earlier.

In the elementary treatment of shot noise, it is always assumed that the instants of departures of electrons from the cathode are distributed according to the Poisson distribution (Chapter 1):

$$P_k(t) = (\sigma t)^k \exp(-\sigma t)/k! \qquad 9.20$$

$P_k(t)$ is the probability of exactly k departures in time t, and σ is the average number of departures per unit time; that is $\sigma = \langle I_0 \rangle / e$, for we now work in the actual current rather than in the current density.

Under the assumptions made about the diode, and with the additional assumption of zero initial velocity, it is easy to show[3] that the passage of a single electron across the diode is represented by a triangular current pulse:

$$i(t) = 2et/\tau^2 \qquad 9.21$$

The current, then, is the superposition of a very large number of such pulses, started at random instants, distributed according to equation 9.20. It is important to reiterate that the pulse shape is independent of the instant the pulse starts. In turn, this means that we can write the autocorrelation function of the current in the form of the autocorrelation function of the pulses (equation 9.21) multiplied by α constant which comes from the process of averaging over the Poisson distribution, for a long time. In Chapter 1, we showed directly that this average is merely σT, in the notation used here, where T is the measurement time. Working per unit time, then, the constant is $\sigma \langle I_0 \rangle / e$, so the autocorrelation function can be written

$$K_I(s) = e^2 \int G(\zeta + s)\, G(\zeta)\, d\zeta \times \langle I_0 \rangle / e$$

$$= e\langle I_0 \rangle \int G(\zeta)\, G(\zeta + s)\, d\zeta \qquad 9.22$$

However, instead of actually evaluating the pulse autocorrelation function, it is far easier to find the spectral density directly. Equation 9.22 is useful only in that it *proves* that the spectral density evaluated for a single pulse has only to be multiplied by $\langle I_0 \rangle / e$ to obtain the spectral density for the complete random sequence of pulses. To complete the details,

$$I(\omega) = \frac{2e}{2\pi\tau^2} \int_0^\tau t \exp(-j\omega t)\, dt$$

$$= \frac{e}{\pi\theta^2} [(\theta \sin\theta + \cos\theta - 1) - j(\sin\theta - \theta\cos\theta)]$$

where $\theta = \omega\tau$ = transit angle.

Then $2\pi |I(\omega)|^2 = 2\pi\, I(\omega) \,.\, I^*(\omega)$

$$= \frac{4e^2}{2\pi\theta^4} [\theta^2 + 2(1 - \cos\theta - \theta\sin\theta)]$$

For all pulses, this becomes

$$2\pi |I(\omega)|^2 = \frac{e}{2\pi} \,.\, \langle I_0 \rangle \,.\, F(\theta) \qquad 9.23$$

where $$F(\theta) = 4\left[\frac{\theta^2 + 2(1 - \cos\theta - \theta\sin\theta)}{\theta^4}\right] \qquad 9.24$$

Then, from the meaning of $|I(\omega)|^2$, we can write the mean square noise currents in df at f as

$$\langle i^2 \rangle = 2e\langle I_0 \rangle F(\theta)\, df \qquad 9.25$$

$F(\theta)$ decreases monotonically from unity to about 0.1 as θ increases from zero to 2π. Thus, equation 9.25 reduces to the Schottky expression†

$$\langle i^2 \rangle = 2e\langle I_0 \rangle\, df \qquad 9.26$$

at *low* frequencies, while for high frequencies it falls off with $F(\theta)$. If we estimate that low frequencies extend out to 0.2π and that $\tau = 10^{-10}$ s, the relevant frequency is 10^9 Hz, a microwave frequency.

Some remarks are in order about the validity of the theory developed above in application to practical devices. First, space-charge effects alter two characteristics of the system: the pulse shape is somewhat modifed, and this is of little importance at low frequencies, but the space charge introduces a connection between the instants of emission and the previously emitted current. More concretely, if an excess of electrons is emitted during one particular interval τ_c, the field is changed so as to reduce emission during the next interval, and correlation effects working over long times are introduced. From the practical point of view this is extremely beneficial, since the noise is much reduced, but the theory is very much complicated. The classical reference is North[4], but, as far as I know, the theory has not been

†Equation 9.26 can be derived directly, if it is assumed that the individual pulses can be represented by delta functions.

reworked with modern techniques. A second factor, which has been neglected, is that the electrons are emitted with random velocities as well as at random times. At low frequencies, according to the central-limit theorem, one would expect an increase in the mean square fluctuation. Experiment shows that any such increase is small, since equation 9.26 has been accurately tested, and this can be explained qualitatively by the simple recognition that changes in initial value change the transit time by small amounts and that these are not shown in l.f. measurements.

Thermionic diodes are technically important in work on noise since they provide accurate noise-current standards. By varying $\langle I_0 \rangle$, the noise amplitude is readily controllable and, moreover, the available noise power is large.

The noise behaviour detailed above also characterizes diode photocells, because it is immaterial whether the electrons enter the system after thermal excitation or after photoexcitation. It also gives a first-order theory of photomultiplier devices if we take the viewpoint that secondary-emission multiplication gives rise to 'electrons' whose charge is ne instead of e, where n is the multiplication ratio at one dynode. relates to the photocurrent, I_{02} to the current after one dynode, etc., the noise current in the input stage is

$$\langle i_1{}^2 \rangle = 2e \langle I_{01} \rangle \Delta f$$

and that in the next stage is

$$\langle i_2{}^2 \rangle = 2ne \langle I_{02} \rangle \Delta f$$

$$\approx 2n^2 e \langle I_{01} \rangle \Delta f$$

instead of $\langle i_2{}^2 \rangle = 2ne \langle I_{01} \rangle \Delta f$, the simple-minded result. Since the signal power also increases as n^2, the signal–noise ratio is nearly constant along the multiplier. Experimentally it is slightly worsened, but the noise figure is quite small. In photodevices, transit times are orders of magnitude longer than in noise diodes and the l.f. region is correspondingly restricted. All these considerations are relevant to our later discussion of optical detection.

9.7 Flicker noise, or $1/f$ noise

An important phenomenon, which has not been discussed so far, is the excess noise which is encountered in active devices when measurements are made at low frequencies, below about 5 kHz. If the noise power produced by a d.c. coupled amplifier in a bandwidth of, for example, 4–5 Hz centred on frequencies below 5 kHz is measured, it is found that the noise increases above the flat value obtained above 5 kHz and at very low frequencies the increase is very marked. Often, the experimental results approximate very closely to a $1/f$ variation for the excess noise. This accounts for the second name, which is actually a misnomer for the reason that, if an f^{-1} variation were really obeyed down to zero frequencies, the noise power would be infinite. We shall therefore use the term 'flicker noise', although this name is historically associated with thermionic devices and not with semiconducting amplifiers which also show the effect.

The physical processes governing the phenomenon are fairly obvious and it is very easy to get a qualitative understanding of the effect, but it is not yet the case that an adequate theory covering a wide range of circumstances can be given.

If we start by thinking of a thermionic noise diode provided with either a tungsten filament or an oxide-coated cathode, we have to recognize that the vacuum conditions when current is being drawn represent some elaborate quasi-equilibrium in which outgassing is countered by sorption of gas on the cooler parts of the envelope and by attachment to the metal electrode surfaces etc. of gas from the walls (even though during manufacture they may have been treated to prevent outgassing, as much as possible). A slightly more careful than usual analysis of the diode behaviour shows that the current drawn depends on the sum of the battery voltage and the contact potential difference. When this value is put in the emission equation, it is found that the anode current depends on the applied p.d. and on the *anode* work function but within broad limits, not on the cathode work function. If now we think of a metal anode which is partially covered with an adsorbed gas layer, we realize that its work function will differ from that of the pure metal. In general, in the vacuum environment of valves with tungsten filaments it will *exceed* the clean value, but in the oxide cathode vicinity evaporated barium usually reduces the effective work function. Thus, ϕ_{anode} depends on the environment as well as on the base metal.

Now, gas atoms or barium atoms on a metal surface are not particularly stable. The question of how long an atom will dwell on a metal is complicated and depends markedly on whether monolayers or considerably thicker layers are being considered; but, in either case, the layer is only in dynamic equilibrium, with desorption and adsorption in approximate long-term balance. Therefore, the effective work function varies with time and, for that matter, with position, since patches of adsorbed atoms are released every now and then. The effective anode potential is therefore subject to a fluctuation which is naturally reflected in the anode current. If the average time constant of the desorption process is τ_d, we should expect the current to vary with frequency according to $(1 + \omega\tau_d)^{-1}$, which over a certain range can give an apparent ω^{-1} variation. So far, so good.

Consider now a semiconducting diode. Here, there are many processes which can give rise to small slow variations in the current; examples are the slow drift of unwanted impurities to the junction region, annealing of the solid-state material due to high-current operation, and so on. Therefore, we are very willing to believe that relaxation processes exist and can give rise to flicker effects, because estimates of the relevant time constants are not utterly inconsistent with the observed frequency spectra. However, it is not the case that more detailed calculations confirm the theory. The basic problem is usually that, although we should expect a range of relaxation times of roughly comparable values and therefore a complicated frequency variation, it is quite normal to find a good linear relationship with an exponent very close to unity, even down to extremely low frequencies. Since it seems very unlikely that a common dominant process swamps all the others over a large range of junction sizes, materials, etc., the position of the theory is far from satisfactory.

9.8 The fluctuation-dissipation theorem: classical and quantum forms

The discussion of the classical form of this important theorem need not detain us very long. We have already seen how the solution of the noise problem in a dissipative circuit using the Langevin equation leads to Nyquist's theorem and therefore relates the fluctuations to the resistance, the dissipative element in the circuit. It is therefore proved that a Langevin equation, which is of the form

$$a\frac{\mathrm{d}x}{\mathrm{d}t} + bx = y(t) \tag{9.27}$$

will give rise to fluctuations which depend on the finite value of b. Thus, all mechanical systems which give rise to equations of this type will give rise to noise. Furthermore, if we use the Hamiltonian form of the equations of motion, second-order systems (L–R–C etc.) can be reduced to equations of the above form. Lastly, using state-space methods, more complicated coupled systems yield matrix generalization of the Langevin equation.

For completeness, we quote the results of the analysis of equation 9.27. The expressions relate to the fluctuations in the frequency range $\mathrm{d}f$ and have been summed over negative frequencies.

$$\langle y^2 \rangle = 4kT \,\mathrm{Re}\, Z(\omega)\, \mathrm{d}f$$

$$\langle x^2 \rangle = 4kT \frac{\mathrm{Re}\, Z(\omega)}{|Z(\omega)|^2}$$

$$\langle \dot{x}^2 \rangle = 4kT \frac{\mathrm{Re}\, Z(\omega)}{\omega^2 |Z(\omega)|^2}\, \mathrm{d}f$$

where $Z(\omega) = \mathrm{j}\omega a + b$, i.e. if we Fourier transform equation 9.27 we get

$$(\mathrm{j}\omega a + b)\, x(\omega) = y(\omega)$$

so that $\quad Z(\omega) = y(\omega)/x(\omega)$

In the electrical case, $Z(\omega) = V(\omega)/I(\omega)$, whence the notation. The autocorrelation function is $K(x)(\tau) = kT \exp(-|\tau| b/a)$.

The quantum-mechanical form was derived by Callen and Welton[5] in 1951. The difference is just what we ought to expect, viz. kT is replaced by

$$h\omega\left\{\tfrac{1}{2} + \frac{1}{\exp(\hbar\omega/kT - 1}\right\}$$

or, as it is usually written in this context, $2kT$ is replaced by $\hbar\omega \coth(\hbar\omega/2kT)$ which tends to $2kT$ as $\hbar\omega \to 0$.

The proof is not mathematically difficult but it does involve a greater knowledge of quantum mechanics than has been assumed in the rest of this book. Readily accessible sources are Landau and Lifschitz[6] and Levitch[7]; however, we can outline the ideas of their proofs. A system of particles in a reservoir is acted on by a time-varying perturbing force, e.g. an electromagnetic field, which gives rise to a time-varying component in the Hamiltonian. This perturbation causes transitions to states of higher energy. Also, there are inverse transitions to the original state, but these do not compensate for the whole of the adsorbed power and there is net absorption, when an average is taken over the ensemble using the Gibbs distribution. This absorption is readily expressible in terms of the imaginary part of the generalized susceptibility. Next, the mean square fluctuation of a generalized coordinate is calculated, is Fourier transformed to a frequency basis, and is similarly ensemble averaged.

The final results are

$$\alpha'' = \frac{\pi}{\hbar}\left[1 - \exp\left(-\frac{\hbar\omega}{kT}\right)\right] \sum_{m,\,n} \rho_n |x_{nm}|^2 \delta(\omega + \omega_{nm}) \qquad 9.28$$

$$\langle x^2 \omega\rangle = \tfrac{1}{2}\left[1 + \exp\left(-\frac{\hbar\omega}{kT}\right)\right] \sum_{m,\,n} \rho_n |x_{nm}|^2 \delta(\omega + \omega_{nm}) \qquad 9.29$$

or $$\langle x^2(\omega)\rangle = \frac{\alpha''\hbar}{2\pi} \coth \frac{\hbar\omega}{2kT} \qquad 9.30$$

where α'' = imaginary part of susceptibility

$\rho_n = \exp[(F - E_n)/kT]$

F = Helmholtz free energy

$|x_{nm}|^2$ = modulus of time-independent matrix element of operator x, belonging to coordinate x.

These remarks bring out some important points. The system of particles interacts with photons, i.e. we are considering particles inside a cavity which is supplied with electromagnetic power and this explains why the quantum modification is essentially a matter of Bose–Einstein statistics. The coupling between the particles and the reservoir is electromagnetic in form, as we readily see if we imagine our particles made into a metallic resistor at an elevated temperature.

9.9 Noise in optical systems

There are many excellent textbooks on the noise behaviour of radio systems and it is therefore not necessary to discuss these aspects here. However, the advent of the laser and of low loss optical fibres has made the development of optical communication systems a possibility and it is interesting to discuss noise in the frequency region for which $\hbar\omega \geqslant kT$, that is in which fluctuation of the zero-point energy becomes important. We shall find that many well-known results of radio-frequency engineering reappear in an only slightly modified form.

The treatment to be given is based on what is still a highly idealized version of the laser characteristics. We assume that a laser produces a completely coherent exponential oscillation and that the noise associated with the oscillation is negligible in comparison with the wanted output. We also assume that single-mode operation can always be achieved when required. Another assumption made is that the laser light propagates *in vacuo* so that the propagation medium is non-dispersive and the phase and group velocities are therefore equal and have the common magnitude c. It would be simple to allow for the change of phase velocity in going over to other media, but the question of dispersion raises difficulties which are not essential in a first treatment of the noise.

In section 9.5 we have already quoted the quantum-mechanical uncertainty of the measurement of the canonical variable q as

$$\langle \Delta q^2 f \rangle = \hbar/4\pi f = \hbar/2\omega$$

We interpret this expression in the following way. The laser produces a noiseless, simple harmonic oscillation which can be characterized by the canonical momentum p and coordinate q. We can measure either p or q or both by appropriate measuring systems so, for concreteness, consider first the measurement of q alone – we shall briefly describe how this is done, after deriving the result. The measurement of q is subject to the uncertainty just quoted, i.e. an ensemble of measurements of q will fluctuate between calculable limits. Therefore, a continuous measurement of q will exhibit random time variations of calculable amplitude. But this is just the fundamental noise in the system. The signal–noise power at input to the system can therefore be written in the form

$$\left(\frac{S}{N}\right)_{\mathrm{q}} = \frac{\langle q^2 \rangle}{\langle \Delta q^2 \rangle}$$

where $\langle q^2 \rangle$ = square of the mean amplitude of q

= perceived power of the laser x constant

Both in quantum theory and in classical theory, the energy of a simple harmonic oscillator is

$$E = \tfrac{1}{2}(p^2 + \omega^2 q^2)$$

and of this the part due to q is

$$E_{\mathrm{q}} = \tfrac{1}{2}\omega^2 q^2$$

If the measurement process takes a time τ, a column of radiation, length l, enters the detector device, where $l = c\tau$. Then, the power entering the detector is $E_{\mathrm{q}}\, c/l$. In k-space (or β-space), each mode corresponds with a frequency range, equation 9.5, $2(l/c)\Delta f$; therefore, the power can be written as

$$P_{\mathrm{s}} = 2E_{\mathrm{q}}\Delta f = \omega^2 q^2 \Delta f$$

Thus, the signal–noise ratio becomes

$$\left(\frac{S}{N}\right)_{\mathrm{q}} = \frac{2P_{\mathrm{s}}}{hf\Delta f} \qquad 9.31$$

A similar calculation made at radio frequencies would have had the maximum available noise power $kT\Delta f$ as the denominator, so that we have exchanged hf for kT, which is what we ought to have expected in the frequency range considered. The result shows that to attain a certain signal-to-noise ratio, much more signal power is required at optical frequencies than at radio frequencies. If we refer to Fig. 9.4, we see that, for optical frequencies in the ordinary visible range, the increase would be about two orders of magnitude. In the context of an optical waveguide system this means that the allowable transmission loss from transmitter to receiver is much smaller than in an r.f. system with identical transmitter power.

The relationship between τ and Δf derived above is closely related to the Fourier spectrum of a pulsed harmonic signal, frequency f_0, which spectrum has zeros at $f_0 - 1/\tau, f_0 + 1/\tau$, i.e. $\Delta f_0 = 2/\tau$. According to the details of the receiver, one utilizes more or less frequency range, but the minimum which will give an adequate rendition of the pulse shape is $\Delta f \approx 1/2\tau$.

The calculation using p alone yields exactly the same result, so it is immaterial whether one measures p or q. If *both* p and q are measured, the signal–noise ratio is halved to

$$\left(\frac{S}{N}\right)_{\mathrm{p,\,q}} = \frac{P_{\mathrm{s}}}{hf\Delta f}$$

The proof is left as an exercise to the reader.

So far, we have said nothing about the way in which the detection system actually works. Laser signals are measured by photocells or by photoconductive devices. both of which may be regarded as devices which convert photons into electrons, with a quantum efficiency which ranges from about 0.1 to 0.8 according to the particular device used. The detector proper can either be used with a linear amplifier of appropriate bandwidth or as a superheterodyne or as a homodyne. In both the latter cases the detector is exposed to the simultaneous action of *two* laser light sources. In the heterodyne, one of the sources is the signal and other is a local oscillator tuned to a somewhat different frequency. The current output from the detector then contains a term at the difference frequency which can be amplified in an intermediate-frequency amplifier before final rectification to the baseband. In principle, there is no difference from radio practice, but the technical problems are much greater. In the homodyne, the original laser signal is split in two and one part is used for the signal path, the other serving as the local oscillator. Analysis shows that baseband signals are obtained in the detector output and that, by varying the phase difference between the signal and the local oscillator, either the p or q component can be picked off. All these systems have various merits and demerits, which are extensively discussed in the technical literature both for radio frequencies and for optical frequencies. Here, we need give only a very general outline of the manner in which the input signal-to-noise ratio is worsened by the detector stage.

The calculation is based on the fact that in any detection system the photodetector output has a d.c. term, in addition to any i.f. or baseband terms. The d.c. is subject to full shot noise and this noise worsens the signal-to-noise ratio. The calculation is approximate and implies that the S/N at input to the photodetector is large, i.e. that the signal power is considerably greater than the threshold value.

Consider a superheterodyne system in which ω_0 is $2\pi \times$ local oscillator frequency; $\omega_s = 2\pi \times$ signal frequency; ϕ_0, ϕ_s the corresponding phases. Then, the net electric field vector at the photodiode surface is of the form $E = E_0 \cos(\omega_0 t + \phi_0) + E_s \cos(\omega_s t + \phi_s)$ and the power is proportional to the square of this. Restricting ourselves to the d.c. and the (wanted) difference frequency components, we get

$$P = P_0 + P_s + 2\sqrt{(P_0 P_s)}\cos[(\omega_0 - \omega_s)t + \phi_0 - \phi_s]$$

or, if as is normal, $P_0 \gg P_s$,

$$P \approx P_0 + 2\sqrt{(P_0 P_s)}\cos[(\omega_0 - \omega_s)t + \phi_0 - \phi_s]$$

We now need a relation between optical power and electric current. An optical power P is equivalent to $n\hbar\omega$ photons of frequency ω, and each photon causes η electrons to cross the diode, where η is the quantum efficiency, a function of ω, in general. Then $i = \eta eP/\hbar\omega$. Thus, we can write the output photocurrent as

$$I = \frac{\eta e}{\hbar\omega}(P_0 + 2\sqrt{(P_0 P_s)}\cos[(\omega_0 - \omega_s)t + \phi_0 - \phi_s])$$

Under our assumptions, $\sqrt{(P_0 P_s)} \ll P_0$, so we have a large standing current with a superposed i.f. of much smaller magnitude. Averaging the cosine term, we find

$$P_{\text{i.f.}} \propto \langle i_{\text{i.f.}}^2\rangle = 2\left[\frac{\eta e}{\hbar\omega}\cdot\sqrt{(P_0 P_s)}\right]^2$$

while the noise associated with the d.c. component is

$$\langle i_n{}^2\rangle = 2e\,I_0\,\Delta f = \frac{2\eta e^2}{\hbar\omega}\,.\,P_0\,\Delta f$$

$$\text{Therefore}\quad \left(\frac{S}{N}\right)_{\text{o/p}} = \frac{\langle i_{\text{i.f.}}^2\rangle}{\langle i_n{}^2\rangle} = \eta h\,\frac{P_s}{f\Delta f}$$

Since $\eta < 1.0$, this means the signal–noise ratio is worsened and explains the importance of high quantum efficiency.

We can take this calculation further by consideration of Fig. 9.5. The photodiode has a conversion 'gain' which is directly proportional to $2(\eta e/\hbar\omega)^2\ .\ P_0$ i.e. to $\langle i_{\text{i.f.}}^2\rangle/P_s$. This gain is called G in the figure, and the output signal power is just GP_s. The diode amplifies the input quantum noise $P_n = hf\,\Delta f$. Also, the diode introduces extra noise equal to

$$P_D = 2(\eta e/hf)\ e\ P_0\,\Delta f$$

Therefore $P_D/G = hf/\eta\Delta f$

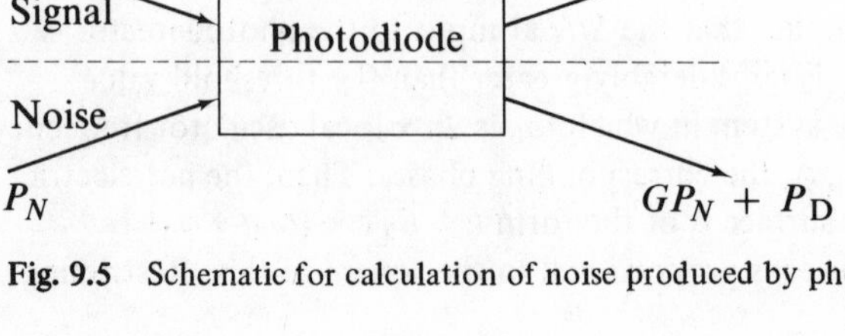

Fig. 9.5 Schematic for calculation of noise produced by photodiode.

The overall signal-to-noise power at the output is

$$\left(\frac{S}{N}\right)_{o/p,\, T} = \frac{GP_s}{GP_n + P_D} = \frac{P_s}{P_n + P_D/G} = \frac{P_s}{hf(1 + 1/\eta)\Delta f}$$

In radio terminology this states that the noise figure of the photodiode is $F = (1 + 1\eta)$. A 'perfect' active device would have $F = 1$. We see that low quantum efficiency gives a very considerable increase in noise figure. Some detectors, photomultipliers and avalanche photodiodes, give large values of G. Our expression then reduces to P_s/P_n, that is, to the ideal quartum limited value.

The computation can now be extended to a chain of amplifiers in the usual manner. Call the noise figure evaluated above F_1 and let the next stage have noise figure F_2. F_2 must be calculated with reference to $hf\Delta f$ and *not* the usual $kT\Delta f$, so it is smaller than the r.f. value. Then, the overall noise figure at the input is

$$F_{12} = F_1 + \frac{F_2 - 1}{G_1} + \text{(factors from subsequent stages.)}$$

where G_1 = available gain of *first* stage. As usual, we must try to make F_1 small and G_1 large to obtain a good overall noise performance.

10

Information Theory and Statistical Mechanics

Many readers of this book will have had, at least, an introductory course in information theory and will have met the concept of the entropy of a source of information. They will probably have been told that the name arose from the, similarly defined, statistical entropy we have used in this book. Also, they may have been exposed to the idea that there is, or may be, a deep connection between the two subjects – information theory and statistical mechanics. It is therefore necessary to discuss the nature of the connection and whether there are grounds for believing that there is more than a fortuitous resemblance between a few formulae.

'Information theory' here is used to mean the study of the methods by which messages (the events which convey information) are coded, decoded, transmitted and contaminated by noise during transmission – not the detailed study of methods of modulation, etc., which is the province of communications theory. At the start, it is to be noted that information theory has registered its major successes in the study of transmission systems in which the signals take the form of groups of pulses. In the older, analogue transmission field, e.g. simple voice telephony and broadcasting of radio or TV, information theory as it now stands has little to offer; we shall therefore, confine our work to pulse systems or, as they are often called, digital systems.

Information theory was really born from the telegraph industry, and significant early contributions were made by R. V. Hartley in the 1920's. However, the subject developed very rapidly after the discoveries of C. E. Shannon in the period 1945–50. Subsequently there was a period of exceedingly active development in which considerable numbers of important problems were solved, but it is perhaps not an insult to information theorists to say that at present the subject is rather dormant and that, at any rate for the moment, more interest is shown in communications theory.

10.1 The Hartley–Shannon measure of information

Before one can talk scientifically about information one has to establish some method of measuring it. Clearly, this is largely a matter of intuition, since in ordinary verbal terms if a statement such as 'my Italian car is black' is made to several persons simultaneously, it may convey quite different degrees of information to the several individuals hearing the remark. To get some sort of measure we have to abstract ourselves from the everyday world and consider the problem of sending such messages over a transmission system. One way of doing this was suggested long ago by S. Morse, who had the seminal idea of coding each letter of the message into an arrangement of short and long pulses which could be transmitted along a pair of wires and decoded. This takes us to the idea that each letter has an information content. However, it is obvious that the information content of letters in English text is by no means the same because, for example, we can entirely leave out some letters, *q* for instance, while other letters are essential. We might, then, go further and conclude that the information content of the letter is really bound up with the number of pulses required to send it.

If we wish to send 64 letter symbols – the alphabet plus digits plus various typographical elements, for example – and we do this by either transmitting a unit pulse or not transmitting anything, then in each time slot there are two possibilities, in n time slots 2^n possibilities and for $n = 6$ we should be able to transmit our 64 symbols. However, this is not yet sufficient because we shall obviously use some symbols much more often than others and we must clearly allow for this in an efficient code. Another point which can now be made is that the information ought to be logarithmic, because adding numbers of time slots multiplies the number of symbols which can be sent. For example, if we operate two six-digit channels side by side, we could send 128 symbols; but if, at the receiver, we add the six digits of one channel in ordered fashion to the digits of the other, we can send $64 \times 64 = 2^{12}$ symbols.

Suppose we now take a six-digit channel sending 64 symbols and, over a long period of normal use, actually count the number of times each symbol is sent. Then, on the frequency definition of probability, we can assign a reasonably accurate value to the probability of each symbol. We can then associate an information measure with the probability in some way. Shannon's choice was to reason that the more improbable the symbol, the greater the information transfer whenever it is actually sent. Some people find this a natural and proper choice, others don't: we shall see that this is not important, as long as we maintain a fixed definition. The Shannon definition of the self-information content of a symbol x_i which occurs with probability p_i is

$$\text{self information} = \log(1/p_i) = -\log p_i \qquad 10.1$$

Thus, the smaller p_i the more improbable is x_i and the larger the information, as was desired.

Obviously for all the symbols $x_1, x_2, \ldots, x_M$ we have, as usual,

$$\sum_{i=1}^{M} p_i = 1.0 \qquad 10.2$$

Also, probability unity or certainty gives zero information and impossibility; probability zero gives infinite information.

Consider now a very large number of symbols N. The expectation values of x_1, x_2, etc. are Np_1, Np_2, etc. and therefore the expected information is $-Np_1 \log p_1$, $-Np_2 \log p_2$, etc. Thus, the average information per symbol is

$$H = \frac{1}{N} \sum_{i=1}^{M} (-p_i \log p_i)\, N$$

$$= -\sum_{i=1}^{M} p_i \log p_i \qquad 10.3$$

Since this is the same as the statistical expression for the entropy according to Boltzmann except for the constant k, H is called the 'entropy' of the 'alphabet' or message source of M symbols.* Thus, we have a basic similarity: in statistical mechanics increased entropy roughly means more disorder, in information theory it measures the uncertainty of the message.

The base of the logarithms is conveniently taken as 2 for (0, 1) binary coding, and we shall subsequently abbreviate $\log_2$ to log, as there will be no confusion with the natural logarithms of the rest of this book. When this is agreed, the units of information defined by equation 10.1 are *bits* and, since H is obtained just by multiplication by a pure number (p_i), bits are also the units of entropy.

The function H has the following properties.

1. $H \geqslant 0$, being zero only if $M - 1$ probabilities are zero and the last is unity.
2. $H \leqslant \log M$; $H = \log M$ if all the probabilities are equal and therefore have the value M^{-1}.

This is readily proved by the following argument. Let all the probabilities $= 1/M$ except for $p_1 = (1 + \epsilon) M^{-1}$ and $p_2 = (1 - \epsilon) M^{-2}$ so that Σp_i is still unity. Then, direct calculation gives

$$H = + \log M[1 - \epsilon^2/2M \log M]$$

so that the entropy is reduced by the error term. The result can be extended by induction.

For our message source, $H = 6.0$ bits. For the English alphabet with no spaces or allowances for punctuation etc. the entropy is 4.7 bits for equal probabilities and about 4.1 bits/letter if the relative frequencies are allowed for. However, for a page of English text, the entropy is much lower because relatively few letter combinations make words and only some word combinations made sense. Shannon has estimated that the overall result is only about 1 bit/letter.

With this degree of understanding of the information concept in communication, we can better appreciate the manner in which Jaynes made use of information entropy to rederive Gibbs' canonical distribution theorem. Jaynes introduced the postulate that, if we take a complex mechanical system about which only partial data are given in the form of a discrete set of probabilities p_i for the several energy states e_i, the optimum manner of guessing (more politely assigning) the p_i's is to choose them so that $\sum_i p_i \ln p_i$ is a minimum. The minimisation must be consistent with any information we do have. We have just seen that the sum above will actually take on a minimum value when all the p_i's are equal, unless further constraints are specified.

In this formulation the basic problem of statistical mechanics becomes the computation of the p_i's so as to minimise $\sum_i p_i \ln p_i$ subject to

$$\sum_i p_i = 1.0$$

$$\sum_i p_i e_i = E_T$$

*Since $p_i < 1.0$, H is a positive number, which is rather confusing.

as usual. As in chapter 3, we vary these expressions and introduce Lagrange multipliers a and β. Then

$$\sum_i (1 + \ln p_i)\delta p_i = 0$$

$$a \sum_i \delta p_i = 0$$

$$\beta \sum_i e_i \, \delta \, p_i = 0$$

$$\text{or} \sum_i (\ln p_i + a + \beta \, e_i) \, \delta \, p_i = 0$$

As before, $p_i = \exp(-a). \exp - (\beta e_i)$

or, because $\sum p_i = 1.0$

$$p_i = \frac{\exp - \beta e_i}{\sum_i \exp - \beta e_i}$$

Notice that Jaynes' postulate is on the probabilities of finding a particle in a given energy state and not with the numbers of particles in a given state. The probabilities can be converted to expectation values by multiplication by a *constant* total number of particles N. Thus, the development is of the canonical rather than the grand canonical distribution.

The reader will find it instructive to compare this section with the development of equation 3.83, which depended on an assumption equivalent to that of equal probabilities of all the states. It is clear that there we worked forward to an expression for the entropy from defined probabilities, here we work back from an assumed entropy expression to the Gibbs distribution. The entropies are the same, except for k and the sign. Both differences are inessential; k specifies the units, the sign is arbitrary. In fact, many writers on statistical mechanics state Jaynes' postulate as a maximisation of $- \sum_i p_i \ln p_i$. It seems slightly more in accordance with verbal usage to say that one should maximise the uncertainty. This is done by the present choice of sign but it has to be admitted that standardization would be welcome.

10.2 Mutual information

So far, we have discussed the information available at the source. What is the situation at the receiver? Fig. 10.1 shows a schematic system. The transmitter informa-

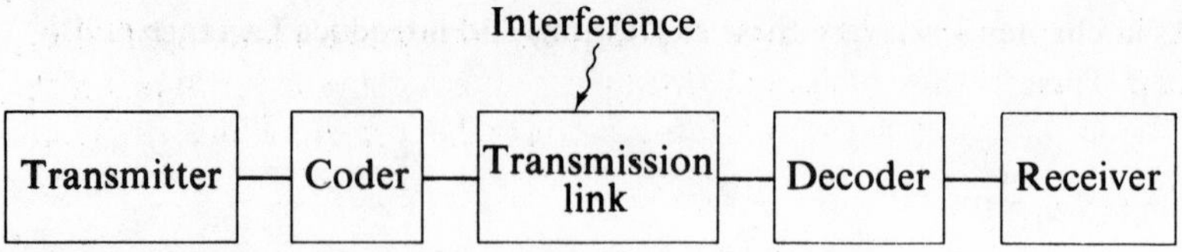

Fig. 10.1 System for information transmission.

tion is coded, sent over a transmission link, decoded and then passed to the receiver. The input signals are therefore processed in ways which will not exactly reproduce their true form, i.e. they are distorted. The coder and decoder may misfunction or be unable to recognize the input signal sufficiently well to treat it as it ought to be. The transmission link is subject to interference of various kinds and the decoder is subject to electrical noise; therefore, it is by no means certain that a particular x will be received as x; instead it may be transposed into one of the remaining $(M-1)$ symbols. If, at the receiver, x_j is actually received, then (Chapter 1) the probability that x_j was sent is a conditional probability. Thus, we can write that, if x_j occurs, $P(x_i/x_j)$ is the probability that x_i was sent. Of course, we try to make all the conditional probabilities as small as possible, except for $P(x_i/x_i)$ which we should like to be unity.

The *mutual information* is defined as

$$I_m(x_i/x_j) = \log\left[\frac{p(x_i/x_j)}{p(x_i)}\right] \qquad 10.4$$

It will be clearer now if we use y_i as the symbol for events at the receiver: we can think of y_i as x_i subject to a scale change. Rewriting, we get

$$I_m(x_i/y_j) = \log\left[\frac{p(x_i/y_j)}{p(x_i)}\right] \qquad 10.5$$

Mutual information has the following properties.

1. Trivially, I_m is a maximum for a conditional probability of unity.
2. I_m is symmetrical in x and y. From equations 1.1 and 1.2, we see that

$$\frac{p(y/x)}{p(x/y)} = \frac{p(y)}{p(x)} \quad \text{or} \quad \frac{p(y/x)}{p(y)} = \frac{p(x/y)}{p(x)}$$

3. If y_i and y_j are received relating to two transmitted symbols, x_k, x_l, the information gain is the sum of the two separate information gains. Since x_k, x_l are independent,

$$p(x_k, x_l) = p(x_k) \cdot p(x_l)$$

$$p(x_k x_l/y_i y_j) = p(x_k/y_i) \cdot p(x_l/y_j)$$

Therefore $$I_m = \log\left[\frac{p(x_k x_l/y_i y_j)}{p(x_k, x_l)}\right]$$

$$= \log\left[\frac{p(x_k/y_i)}{p(x_k)}\right] + \log\left[\frac{p(x_l/y_j)}{p(x_l)}\right]$$

4. If y_i and y_j are received in succession, relating to *one* transmitted symbol x_k, and the receiver treats the conditional probability of the first as the (*a priori*) probability of the second, then the mutual information is the sum of the two separate values.

By definition, $$I_m = \log\left[\frac{p(x_k/y_i y_j)}{p(x_k)}\right]$$

$$= \log\left[\frac{p(x_k/y_i y_j)}{p(x_k/y_i)} \cdot \frac{p(x_k/y_i)}{p(x_k)}\right]$$

$$= \log\left[\frac{p(x_k/y_i y_j)}{p(x_k/y_i)}\right] + \log\left[\frac{p(x_k/y_i)}{p(x_k)}\right]$$

Next, we define the average mutual information. This is done with respect to the joint probability $p(x_i \,.\, x_j)$ and this is

$$I_m(X, Y) = \sum_i \sum_j p(x_i, y_j) \log\left[\frac{p(x_i/y_j)}{p(x_i)}\right] \qquad 10.6$$

Two entropies are also defined. The first is the system entropy

$$H(X, Y) = -\sum_i \sum_j p(x_i, y_j) \log p(x_i y_j)$$

which is

$$H(X, Y) = H(X) + H(Y) \qquad 10.7$$

if the X's and Y's are independent, i.e. if $p(x/y) = p(x) \,.\, p(y)$.

Secondly,

$$H(X/Y) = -\sum_i \sum_j p(x_i, y_j) \log p(y_j/x_i) \qquad 10.8$$

If $p(x/y) \neq p(x) \,.\, p(y)$, the relation for $H(X,Y)$ becomes

$$H(X, Y) = H(X) + H(Y/X) \qquad 10.9$$

$$= H(Y) + H(X/Y) \qquad 10.10$$

Then, it is easy to show by expansion that

$$I_m(X, Y) = H(X) + H(Y) - H(X, Y) \qquad 10.11$$

If we eliminate $H(Y)$ from equation 10.11, using equation 10.10 the result is

$$I_m(X, Y) = H(X) - H(X/Y) \qquad 10.12$$

This equation has an important and simple physical meaning. $H(X)$ is the entropy content of the source, therefore $H(X/Y)$ represents the information *lost* in transmission. $H(X/Y)$ is termed the 'equivocation' of the circuit.

In turn, if we eliminate $H(X)$, using equation 10.9 we find

$$I_m(X, Y) = H(Y) - H(Y/X) \tag{10.13}$$

Since $H(Y)$ is the entropy at the receiver, we can look on $H(Y/X)$ as an entropy loss in reception or in noise; $H(Y/X)$ is therefore often called the 'noise entropy' of the system.

10.3 Continuous information source

We have already said that information theory is better adapted to dealing with digital signals rather with than analogue signals. However, the latter are very important, and some progress can be made with them by translating them into pulses.

The sampling theorem states that if a waveform contains no frequencies in excess of some fixed value F, then it can be reconstructed from a series of amplitude measurements made at intervals of $(2F)^{-1}$ seconds. Actually, some tolerance has to be allowed and a slightly greater sampling rate must be used. Thus, the output of a simple sampler consists of a regular series of pulses whose amplitudes are distributed

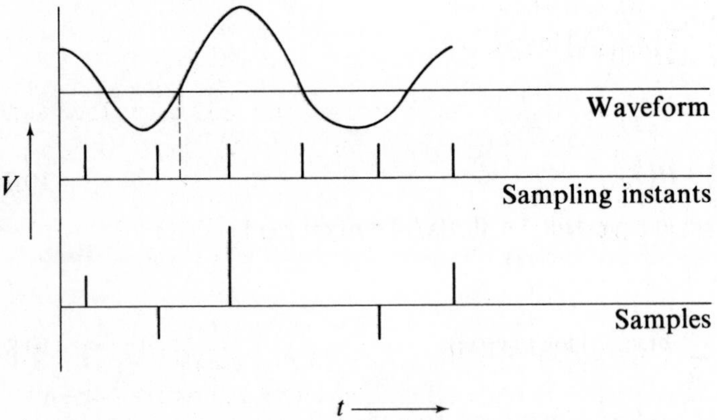

Fig. 10.2 To illustrate sampling and the sampling theorem.

according to the input signal. Fig. 10.2 illustrates the principle. The probability of a sample having an amplitude in dv at v is $p(v_i)\,dv_i$ and the corresponding entropy for each sample is

$$H = -\sum_i p(v_i)dv_i \,.\, \log\,[p(v_i)dv_i]$$

$$= -\sum_i p(v_i)\log[p(v_i)]\;dv_i - \sum_i p(v_i)\log[dv_i]\;dv_i$$

or, on taking the limit as $dv_i \to 0$,

$$H = -\int_i p(v_i) \log[p(v_i)] \, dv_i - \infty \qquad 10.14$$

The infinite constant is not a worry because, as in classical thermodynamics, we normally work in entropy differences; but there are other difficulties about equation

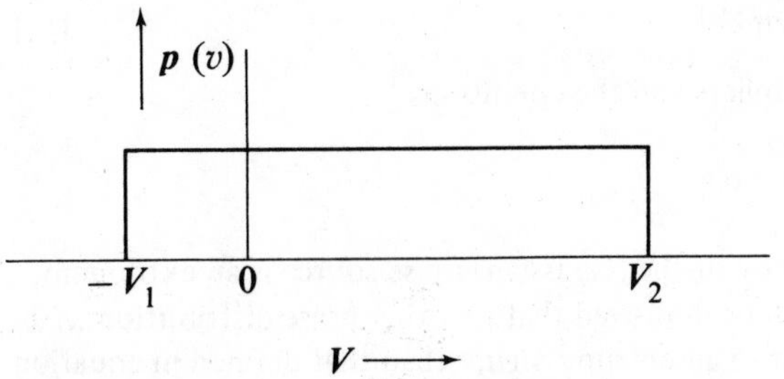

Fig. 10.3 The sign of the information entropy of this distribution depends on whether $V_1 + V_2$ is greater or less than unity.

10.14. First, it gives values of H which can be positive, zero or negative. The standard illustration is Fig. 10.3, a uniform distribution from $-v_1$ to v_2.

$$H = -\log \frac{1}{v_1 + v_2} \cdot \frac{1}{v_1 + v_2} [v]_{v_1}^{v_2}$$

$$= \log(v_1 + v_2)$$

Thus H is positive if $(v_1 + v_2) > 1.0$; zero if $(v_1 + v_2) = 1.0$; negative if $(v_1 + v_2) < 1.0$.

Notice that as long as $(v_1 + v_2)$ = constant it is immaterial where the range is taken; thus, equation 10.14 does not depend on the origin.

If $(v_1 + v_2)$ in the receiver differs from the transmitted value, there is an apparent change of entropy. Investigation shows that the difference comes only in the relative entropy and is taken into the subtractive constant. Thus, care should be taken with this concept, but in the majority of cases we may use it.

An interesting and useful result is obtained when we apply this theory to noise. Consider that a noise voltage has been passed through a filter which has restricted the frequency range to that between 0 and F. Then, the mean noise power is a constant kTF which we call σ^2 for the moment. The corresponding probability density is

$$p(v) = \frac{1}{\sqrt{(2\pi)}\sigma} \cdot \exp\left(\frac{-v^2}{2\sigma^2}\right)$$

Owing to the rapid increase of the exponential, we can take the integration limits as $\pm\infty$, so H becomes*

$$-\int_{-\infty}^{+\infty} p(v)\ln p(v)\,dv = -\langle \ln p(v)\rangle = \ln\sqrt{(2\pi)}\,\sigma + \tfrac{1}{2} = \tfrac{1}{2}(\ln 2\pi e\,.\,\sigma^2)$$

or, explicitly for the noise,

$$H = \tfrac{1}{2}\ln(2\pi e\,.\,P_N) \text{ nats}$$
$$= \tfrac{1}{2}\log(2\pi e\,.\,P_N) \text{ bits}$$
$$= \tfrac{1}{2}\log(2\pi e\,.\,kTF) \text{ bits/symbol.} \qquad 10.15$$

By using Langrangian multipliers and the conditions

$$\int p\,(v)\,dv = 1, \quad \int v^2 p(v)\,dv = \sigma^2$$

it is easy to show that the entropy of the (Gaussian) noise source is an extremum, actually a maximum. It has then been proved that no other noise distribution with the *same* mean power will result in an entropy *higher* than that defined in equation 10.15. Therefore, the noise amplitude distribution or, more generally, the fluctuation amplitude distribution, which is characteristic of statistical mechanical calculations, gives rise to the maximum communication entropy. In one sense, this is not surprising – we have, after all, defined communication entropy in exactly the same mathematical form as statistical entropy and we should therefore hope that consistent application of the theories should lead to the same results. However, it does suggest that the reasons for defining communication entropy as Shannon did are rather more fundamental than at first sight they appear. We shall see the question more clearly when we have considered the actual influence of noise on the information capabilities of a system.

10.4 The noisy channel

In information theory, a channel is best visualized as a pair of wires or a coaxial cable connecting the coder and decoder. Only a certain band of frequencies is transmitted, filters usually being incorporated to remove unwanted frequencies, so the channel bandwidth may be much less than any natural bandwidth of the coaxial cable. For example, dispersion limits the bandwidth of small coaxial cables to the range 0–0(10 MHz), but a telephone channel is limited to 3.5 kHz range within the larger range. The bandwidth fixes the length of the shortest pulse which can be transmitted without excessive distortion according to $\Delta f = 0(1/\tau)$, where τ = pulse length, but with more generality we can *define* the channel capacity (nothing to do with electrostatic capacity) as

$$C = n \log k \text{ bits/second} \qquad 10.16$$

*It is easier to do analytical work with natural logs; numerical calculations are easier with logs to base 2.

if the channel will transmit n pulses of k different amplitudes per second. For example, the channel can be operated with $k = 0, 1$ or with $k = 1, 0, -1; 1, \frac{1}{2}, 0, -\frac{1}{2}, -1$; etc. Theoretically, more levels can be used, but there is a limit due to noise because, eventually, the spacing between levels will become small and, when it is of the order of magnitude of the channel noise, further subdivision is useless. This is the basic reason why noise influences the channel capacity.

To transmit all the possible information from a message source, C must exceed H, but in an efficient system it will only slightly exceed it. Adjusting the information source so as to work efficiently with a given channel gives scope for improved methods of coding, designed so as to reduce the redundancy of the (modified) source. For example, we have said that the English alphabet is equivalent to 4.7 bits while the real average information per symbol is only about 1 bit; therefore a highly efficient coding system could process the messages in such a way that the channel capacity required be reduced by a factor of 4.7. The principle used in such coding is that improbable events be assigned the longer code signals. Coding methods are not germane to our major theme and will not be discussed further.

We are now able to work out the channel capacity for a noisy channel. We are still working with signals (or noise) sampled at the rate $2F$ per second. From equation 10.13, the received information is

$$I_m(X, Y) = H(Y) - H(Y/X)$$

The *maximum* signal entropy will, as before, correspond with a Gaussian mean power distribution, so we can write an equation similar to equation 10.15 for $H(Y)$, i.e.

$$H(Y) = F \log\{2\pi e(P_s + P_N)\}$$

and, in view of the meaning of $H(Y/X)$,

$$H(Y/X) = F \log(2\pi e \,.\, P_N)$$

therefore $\quad H(Y) - H(Y/X) = F \log(1 + P_s/P_N)$ bits/second

so, for a 'matched' channel, one which is exactly capable of transmitting this information,

$$C = F \log(1 + P_s/P_N) \tag{10.17}$$

Thus, the bigger P_s, the greater the effective channel capacity becomes. This celebrated result is known as 'Shannon's second theorem'. It plays the important role of setting a limit on which it is not possible to improve; it does not, of course, mean that any particular coding system and channel will approach the limit given by equation 10.17. One important consequence of equation 10.17 is that the bandwidth of the channel, the time required for a transmission, and the signal-to-noise power ratio are all interchangeable. This follows since, if we require to convey a total information I in time T, we have

$$I = TF \log (1 + P_s/P_N) \tag{10.18}$$

The use of narrow-band circuits for a long time will introduce delay into the system, which may or may not be tolerable according to the use to which the system is put.

Once more, we can observe some analogies between information theory and statistical mechanics. Statistical mechanics tells us that the distribution of fluctuations (noise) really is Gaussian. Information theory exploits this to say that the power of the signal should be distributed in the same way, to maximize the entropy. In communications, however, we can influence the distribution of the signal power, which is rarely the case in statistical physics. From these facts we can derive an important theorem giving the performance limit.

10.5 Phase-space representation of signals and noise

We can carry the analogy a little further. In time T, we have to transmit $2TF$ samples of each signal. The amplitude of the signal at each sampling point is a stochastic variable and we can represent the signal as a point in phase space by plotting the $2TF$ amplitudes on coordinates in phase space. Clearly, the space will be of $n = 2TF$ dimensions to represent the signal. The representative point for subsequent intervals T_1, T_2, T_3 will follow a phase trajectory as though we were following a gas molecule. However, if the sample represents a wanted signal, the phase trajectory will exhibit some deterministic features.

Let the amplitudes of the samples be $x_1, x_2, \ldots, x_n$. Since these samples form a stochastic process, we can form the power merely by summing their squares (there are no coherencies). Then

$$P_s \propto \frac{1}{2TF}\sum_{i=1}^{2TF} x_i^2 = \frac{1}{2TF}\sum_{i=1}^{2TF} x_i^2 \qquad 10.19$$

where we have put the constant equal to unity because it will turn out to be irrelevant. But the representative point must lie on the surface of a hypersphere whose radius is

$$r^2 = \sum_{i=1}^{2TF} x_i^2 \qquad 10.20$$

$$\therefore\ r^2 = 2TFP_s \qquad 10.21$$

Now, contaminate the signal with noise. Clearly

$$r_1^2 = 2TF(P_s + P_N) \qquad 10.22$$

But, we can also regard the noise as moving the signal point to another point inside the sphere of radius

$$r_2^2 = 2TFP_N \qquad 10.23$$

as illustrated, schematically, in Fig. 10.4. The received signal point S can now lie anywhere in the shaded area, but points in the noise sphere outside the main sphere radius r_1, correspond with a total power which exceeds $P_s + P_N$.

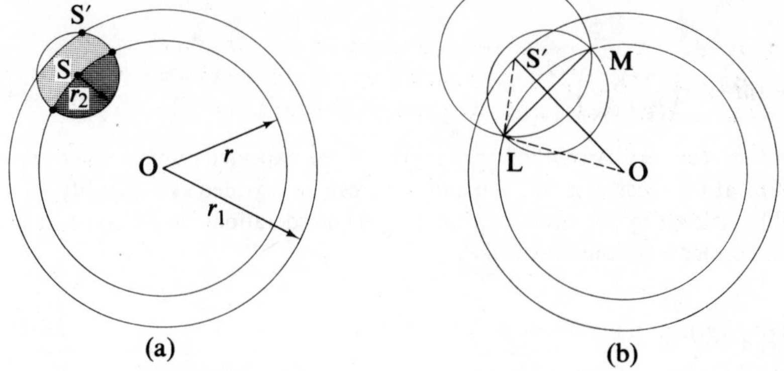

Fig. 10.4 To illustrate the calculation of the relationship between signal power, noise power and error probability.

So far, this is merely a rather interesting representation, but we shall now prove, even if rather informally, that it leads to Shannon's theorem and a slight further advance. From Fig. 10.4, we observe that the *transmitted* signal must have been within the volume common to the spheres of radius r and r_2, double-hatched in 10.4. Then, if transmitted points are distributed such that, on the average, only one point lies within this volume, the receiver can theoretically be made to interpret correctly the received point as belonging to a transmitted point in this volume. The probability of finding one of M representative points in this volume is the cross-hatched volume divided by the volume of the whole hypersphere; call this W.

The volume of an n-dimensional hypersphere is (Chapter 3) $V = \pi^{n/2} r^n/\Gamma(n/2 + 1)$, so the ratio of the volumes of the two hyperspheres, radii r_3, r_4, is just $(r_3/r_4)^n$, confirming that the constants are irrelevant. Fig. 10.4(b) shows the situation, viewed from a possible signal point S′. We make the hypothesis that the cross-hatched volume is less than the volume of the hypersphere with LM as diameter in Fig. 10.4(b).

A little obvious geometry shows that OLS′ is a right-angled triangle and that

$$\frac{\text{LM}}{2} = \frac{\text{OL}\,.\,\text{LS}'}{\text{OS}'} = \left(\frac{2TFP_s\,.\,P_N}{P_s + P_N}\right)$$

$$\text{Therefore} \quad W < \left(\frac{\text{LM}/2}{\text{OL}}\right) < \left(\frac{P_N}{P_s + P_N}\right)^{n/2}$$

The probability that all of the remaining $M - 1$ signal points are *outside* the volume centred on S is $q = (1 - W)^{M-1}$; therefore, for the cross-hatched volume of 10.4(a), we can write the inequality

$$q > \left\{1 - \left(\frac{P_N}{P_s + P_N}\right)^{n/2}\right\}^{M-1}$$

Since $n/2$ is normally a large number, even if P_N is only slightly smaller than P_s, the term $[P_N/(P_N + P_s)]^{n/2}$ is very small and we can use the binomial expansion. Therefore

$$q > 1 - (M-1)\left(\frac{P_N}{P_s + P_N}\right)^{n/2} \tag{10.24}$$

If $q = 1 - \delta$, then, in a large number of signals N, the expected number of errors in translation at the receiver is $N\delta$. The numeric δ can be regarded as a quantity which will be defined by the overall system-design considerations, so we consider it as a known constant. We can now make

$$\delta = M\left(\frac{P_N}{P_s + P_N}\right)^{n/2} \tag{10.25}$$

which satisfies the inequality in equation 10.24, to get

$$M_{\text{max.}} = \delta\left(\frac{P_s + P_N}{P_N}\right)^{n/2} \tag{10.26}$$

for the maximum number of representative points in the hypersphere. Each point, it will be remembered, represents a signal of duration T, and all these signals are to be regarded as having the same probability, i.e. the information per signal is $\log M$ bits, from property 2 of the self entropy. This in turn means that the information rate is $(1/\tau)\log M$ bits/second. Using equation 10.26, the maximum information rate is

$$I_{\text{max.}} = \frac{1}{\tau}\left[\log\delta + TF\log\left(1 + \frac{P_s}{P_N}\right)\right] \tag{10.27}$$

Taking the limit as $T \to \infty$, the maximum information rate $\to F\log(1 + P_s/P_N)$ bits/second, independent of δ.

So we have retrieved equation 10.18 with some extra insight, viz. that equation 10.18 represents a limit which, for a given error rate, can be approached only by increasing the time over which samples are taken. (Remember that $\log\delta$ in equation 10.27 is negative, since $\delta < 1.0$.) The introduction of phase-space ideas has proved useful.

Conclusion

This brief and simple account of information theory serves at least two useful purposes. First, as said at the outset, it shows the reasons why some authors believe that there is a deep significance in the formal similarities between statistical mechanics and information theory. My personal view is that the similarities are interesting and well worth understanding, but do not extend far enough for there to be new possibilities for cross-fertilization between the two disciplines. In particular, I cannot see that information theory has much bearing on kinetic theory, which I regard as the main growth field in modern statistical mechanics. However, several very eminent authors completely disagree and, on grounds which are more philosophical than scientific, believe that a great fundamental truth is involved. Perhaps this difference reflects to a large extent the philosophical importance one attaches to the entropy concept. To those who think of entropy in the terms of the great fathers of thermodynamics, Kelvin and Clausius, information theory is likely to seem more fundamental than to those who remark the fact that, cosmologically speaking, the conservation of entropy cannot operate and are therefore sceptical about the limits of the concept.

The second purpose served concerns the actual context of measurement systems. In research, a great deal of time is often wasted in designing and making inefficient measurement systems. Information theory tells one what sort of procedure is good and what sort is bad, and so I hope that non-engineers will gain some new insights into how their measurement problems can be tackled.

Appendix 1. Mathematical Formulae

Integrals

Several special integrals are important in the text. They start from the Poisson integral:

$$I = \int_{-\infty}^{+\infty} \exp(-\alpha x^2)\, dx = \sqrt{(\pi/\alpha)} \qquad \text{A1.1}$$

The proof follows from expressing the square of the integral in polar coordinates.

Integrals of the even form

$$I_{(2n)} = \int_{-\infty}^{+\infty} x^{2n} \exp(-\alpha x^2)\, dx$$

can be evaluated by differentiation in α and integration by parts. For example,

$$\int_0^{\infty} x^{2n} \exp(-\alpha x^2)\, dx = -\frac{1}{2\alpha}\int_0^{\infty} x^{2n-1}\, d(e^{-\alpha x^2})$$

$$= -\frac{1}{2\alpha}\left[x^{2n-1}\, e^{-\alpha x^2}\right]_0^{\infty} + \frac{2n-1}{2\alpha}\int_0^{\infty} x^{2n-2}\, e^{-\alpha x^2}\, dx$$

leading to the result, *for limits* $\pm\infty$,

$$I_{(2n)} = \frac{2n-1}{2\alpha} \cdot I_{(2n-2)} \qquad \text{A1.2}$$

$$I_2 = \tfrac{1}{2}\sqrt{(\pi/\alpha^3)}$$

$$I_4 = \tfrac{3}{4}\sqrt{(\pi/\alpha^5)}$$

For the odd integrals,

$$I_{(2n+1)} = \int_0^{\infty} x^{2n+1}\, e^{-\alpha x^2}\, dx = n!/2\alpha^{n+1} \qquad \text{A1.3}$$

$$I_1 = 1/2\alpha$$

$$I_3 = 1/2\alpha^2$$

The partition function as a Laplace transform

In the classical case, we have defined the partition function by equation 3.34:

$$Z = \int_0^\infty \exp(-\beta E)\, \Omega(E)\, \mathrm{d}E \qquad \text{A1.4}$$

It is to be remembered that E cannot take on negative values and β is real. Then equation A1.4 is the conventional definition of the Laplace transform of $\Omega(E)$, which can be written down by inversion as

$$\Omega(E) = \frac{1}{2\pi \mathrm{j}} \int_{\beta - \mathrm{j}\infty}^{\beta + \mathrm{j}\infty} Z \exp(\beta E)\, \mathrm{d}\beta \qquad \text{A1.5}$$

where the integration contour is a straight line parallel to the Im(β) axis, with Re(β) chosen so that convergence is assured.

The importance of this observation lies in the fact that a very large body of knowledge of the general properties of Laplace transforms is available; therefore, this can be used to derive results for partition functions, which are very useful.

A simple first example is the partition function for two systems, one with $\Omega_1(E)$ states, the other with $\Omega_2(E)$. The total number states is proportional to the product of $\Omega_1\,\Omega_2$ and can be written as

$$\int \Omega_1(E')\, \Omega_2(E - E')\, \mathrm{d}E \qquad \text{A1.6}$$

or $$\int \Omega_1(E - E'')\, \Omega_2(E'')\, \mathrm{d}E \qquad \text{A1.7}$$

which are, by definition, convolution products. But the Laplace transform rule is

$$L[f * g] = L[f] \,.\, L[g] \qquad \text{A1.8}$$

where * denotes convolution and the r.h.s. is a normal product. Thus, equation A1.8 constitutes a proof that

$$Z(\beta)_{1,2} = Z(\beta_1) \,.\, Z(\beta_2) \qquad \text{A1.9}$$

By induction, this result can be extended to more systems.

Appendix 2. Some Thermodynamic Relationships

A few thermodynamic relationships which much simplify some calculations are given below. The first set involves the free energy F_H, the entropy, the pressure, and the temperature.

From equation 3.47(b) we have

$$dE = TdS - pdV$$

$$= d(TS) - SdT - pdV$$

But $F_H = E - TS$ by definition; therefore

$$dF_H = -SdT - pdV \qquad \text{A2.1}$$

However, F_H is a function of T and V, and so we can write

$$dF_H = (\partial F_H/\partial T)_V \, dT + (\partial F_H/\partial V)_T \, dV \qquad \text{A2.2}$$

Identifying terms in equation A2.1, we get

$$S = -(\partial F_H/\partial T)_V \qquad \text{A2.3}$$

$$p = -(\partial F_H/\partial V)_T \qquad \text{A2.4}$$

as new expressions for S and p.

We can write $E = F - T(\partial F_H/\partial T)_V$ by using equation A2.3 in the definition of F. This can be more usefully applied in the form

$$\left\{\frac{\partial}{\partial T}\left(\frac{F_H}{T}\right)\right\}_V = \frac{1}{T}\left(\frac{\partial E_H}{\partial T}\right)_V - \frac{F_H}{T^2}$$

$$= \frac{E}{T^2} \qquad \text{A2.5}$$

Similar results can be obtained from using the Gibbs' free energy G, which is a function of T and pressure p, defined as

$$dG = -S\,dT + Vdp \qquad \text{A2.6}$$

Analogies of A2.3 and A2.4 are

$$S = -(\partial G/\partial T)_p \qquad \text{A2.7}$$

$$V = (\partial G/\partial p)_S \qquad \text{A2.8}$$

The four relationships, equations A2.3, A2.4, A2.7, and A2.8, show that S, p and V have the nature of generalized forces resulting from *potentials* F_H and G, which explains why F_H and G are called 'thermodynamical potentials' in textbooks on the subject.

To complete the permutations of S, p and V, a third potential, called the 'enthalpy', is introduced. We do not require it in the text, but it is defined by

$$H = E + pV \tag{A2.9}$$

and $$T = (\partial H/\partial S)_p \tag{A2.10}$$

$$V = (\partial H/\partial S)_S \tag{A2.11}$$

Appendix 3. The Evaluation of the Fowler Integral

The integral to be evaluated is

$$\int_0^\infty dy \,.\, \ln\{1 + \exp[\hbar(\omega - \omega_0)/kT - g]\}$$

Let $\hbar(\omega - \omega_0)/kT = \epsilon$. When ϵ is negative, that is $\omega < \omega_0$, $\exp(\epsilon - y) < 1$ and the ln term can be expanded in series. Then

$$I_1 = \sum_{n=1}^{\infty} (-1)^{n-1} \frac{\exp n\epsilon}{n} \int_0^\infty \exp(-ny)\, dy$$

$$= \sum_{n=1}^{\infty} \frac{(-1)^{n-1} \exp n\epsilon}{n^2} = F(\epsilon), \text{ say} \qquad \text{A3.1}$$

When ϵ is positive, $\omega > \omega_0$. Divide the range in two, so that

$$I_2 = \int_0^\epsilon + \int_\epsilon^\infty$$

Let $x = \epsilon - y$. The first integral becomes

$$\int_0^\epsilon dx \ln(1 + \exp x) = \int_0^\epsilon dx \ln[\exp x\{1 - \exp(-x)\}]$$

$$= \int_0^\epsilon dx\, [x + \ln\{1 - \exp(-x)\}] = \epsilon^2/2 + \int_0^\epsilon \ln\{1 - \exp(-x)\} dx$$

As in the derivation of equation A3.1,

$$\int_0^\epsilon \{\ln 1 - \exp(-x)\}\, dx = \sum_{n=1}^{\infty} \frac{(-1)^n}{n} \int_0^\epsilon \exp(-nx)\, dx$$

$$= \sum_{n=1}^{\infty} \frac{(-1)^{n-1}}{n^2} \{1 - \exp(-n\epsilon)\}$$

$$= F(O) - F(-\epsilon) \qquad \text{A3.2}$$

when we exploit the similarity to equation A3.1.

The second integral, which is the same as

$\int_0^\infty dx \ln\{1+\exp(-x)\}$, integrates by parts to

$$\Big[x\ln\{1+\exp(-x)\}\Big]_0^\infty + \int_0^\infty \frac{x\,dx}{1+\exp x} = \int_0^\infty \frac{x\,dx}{\exp x\{1+\exp(-x)\}}$$

Let $\exp(-x) = t$; then

$$\int = -\int_0^1 \frac{\ln t}{1+t}\,dt = \frac{\pi^2}{12}$$

But we also observe that this integral is $F(O)$, by definition. Collecting results, we get

$$I_2 = \epsilon^2/6 + \pi^2/6 - F(-\epsilon) \qquad \text{A3.3}$$

Thus, when $\epsilon = 0$, $I_2 = \pi^2/12 = I_1$. This means that, at the Einstein cut-off frequency, the accurate theory gives a substantial photoemission. For large ϵ, $\epsilon > 10$, $I_2 \approx \epsilon^2/2$ so there is a variation of about two orders of magnitude in the emission from $\epsilon = 0$ to $\epsilon = 10$.

Appendix 4. Derivation of the Fokker–Planck Equation

A neat derivation of the one-dimensional form of the Fokker–Planck equation is given by Levich[1]. The three-dimensional form is discussed by Chandrasekhar[2].

Levich multiplies both sides of equation 6.52 by $\phi(\lambda)/\Delta t$ and integrates over λ. The term $\phi(\lambda)$ is continuous and $\to 0$ as $\lambda \to \pm\infty$, but it is otherwise arbitrary and will finally drop out of the calculation. Then

$$\frac{1}{\Delta t}\int p(\lambda, t + \Delta t)\, \phi(\lambda)\, d\lambda = \frac{1}{\Delta t}\int p(\lambda_0, t)\, d\lambda_0 \int W(\lambda_0, \lambda, \Delta t)\, \phi(\lambda)\, d\lambda \qquad \text{A4.1}$$

Expanding ϕ in a Taylor series in $\lambda - \lambda_0$ and inserting into equation A4.1,

$$\phi = \phi(\lambda_0) + \phi'(\lambda - \lambda_0) + \frac{\phi''}{2}(\lambda - \lambda_0)^2$$

Then, the r.h.s. of equation A4.1 is

$$\int p(\lambda_0, t)\, d\lambda_0 \int \left\{ \phi(\lambda_0) + \phi'(\lambda - \lambda_0) + \frac{\phi''}{2}(\lambda - \lambda_0)^2 \right\} W(\lambda_0, \lambda, \Delta t)\, d\lambda$$

Consider the three separate integrals of the inner integration in turn. In the first, $\phi(\lambda_0)$ can be taken outside the integral, so the result is unity.

The second is

$$\phi' \int (\lambda - \lambda_0)\, W(\lambda_0, \lambda, \Delta t)\, d\lambda = \phi'\, I_1(\lambda_0)$$

and the third is

$$\frac{\phi''}{2} \int (\lambda - \lambda_0)^2\, W(\lambda_0, \lambda, \Delta t)\, d\lambda = \phi''\, I_2(\lambda_0)$$

Since W is a rapidly diminishing function of $\lambda - \lambda_0$, the I_n are finite and rapidly convergent; also, our three-term approximation to ϕ should be quite adequate. We are now left with the following approximation for equation A4.1:

$$\frac{1}{\Delta t}\int p(\lambda, t + \Delta t)\, \phi(\lambda)\, d\lambda - \frac{1}{\Delta t}\int p(\lambda_0, t)\, \phi(\lambda_0)\, d\lambda_0$$

$$= \frac{1}{\Delta t}\int p(\lambda_0, t)\, \phi'(\lambda_0)\, I_1(\lambda_0)\, d\lambda_0 + \frac{1}{2\Delta t}\int p(\lambda_0, t)\, \phi''(\lambda_0)\, I_2(\lambda_0)\, d\lambda_0$$

If we now take the second integral on the l.h.s. over λ and let $\Delta t \to 0$, the result is

$$\int \frac{\partial}{\partial t} p(\lambda, t) \,.\, \phi(\lambda)\, d\lambda = \int p(\lambda_0, t)\, \phi'(\lambda_0) \left\{ \lim_{\Delta t \to 0} \frac{I_1(\lambda_0, t)}{\Delta t} \right\} d\lambda_0$$

$$+ \tfrac{1}{2} \int p(\lambda_0, t)\, \phi''(\lambda_0) \left\{ \lim_{\Delta t \to 0} \frac{I_2(\lambda_0, t)}{\Delta t} \right\} d\lambda_0$$

Let $\left\{ \lim_{\Delta t \to 0} \frac{I_1(\lambda_0, t)}{\Delta t} \right\} = a(\lambda_0)$

$$\left\{ \lim_{\Delta t \to 0} \frac{I_2(\lambda_0 t)}{\Delta t} \right\} = D(\lambda_0)$$

and let $\lambda_0 \to \lambda$ on the r.h.s. The result is

$$\int \frac{\partial}{\partial t}\, p(\lambda, t) \,.\, \phi(\lambda)\, d\lambda = \int p(\lambda, t)\, \phi'(\lambda)\, a\,(\lambda) d\lambda + \int p(\lambda, t)\, \phi''(\lambda)\, D(\lambda)\, d\lambda$$

The two integrals on the r.h.s. can be simplified by partial integration, remembering that $\phi(\lambda) \to 0$ at $\pm\infty$. The results are

$$\int p(\lambda, t)\, \phi'(\lambda)\, a(\lambda)\, d\lambda = -\int \phi(\lambda) \frac{\partial(ap)}{\partial \lambda} \,.\, d\lambda$$

$$\int p(\lambda, t)\, \phi''(\lambda)\, D(\lambda)\, d\lambda = \int \phi(\lambda) \frac{\partial^2(Dp)}{\partial \lambda^2}\, d\lambda$$

Then $\displaystyle\int \left(\frac{\partial p}{\partial t} + \frac{\partial(ap)}{\partial \lambda} - \frac{\partial^2(Dp)}{\partial \lambda^2} \right) \phi(\lambda)\, d\lambda = 0$

This can be satisfied for arbitrary $\phi(\lambda)$ only if

$$\frac{\partial p}{\partial t} + \frac{\partial(ap)}{\partial \lambda} - \frac{\partial^2(Dp)}{\partial \lambda^2} = 0$$

which is the form quoted in equation 6.53.

The reader will have noticed that there is a sign change in the second term, with respect to Einstein's equation, equation 6.50. This comes from the definition of a as a function of $\lambda - \lambda_0$. It could equally well have been defined as $-(\lambda_0 - \lambda)$ without altering the sign of D, and this would make the two forms identical. Therefore, the sign is not relevant, so long as the various definitions are kept consistent.

Appendix 5. The Wiener–Khinchin Theorem

The autocorrelation function $K(\tau)$ is defined as $K(\tau) = \langle f(t) \, . \, f(t+\tau)\rangle$, where time averaging is meant. However, in time-varying fluctuation theory we must impose the condition on the variations that their mean square amplitude is numerically equal to the amplitude of the fluctuations, calculated from ensemble theory; thus, it is immaterial whether time averaging or ensemble averaging is used.

$K(\tau)$ is symmetrical and real, and thus we have

$$K(\tau) = K(-\tau) = K^*(\tau) = K^*(-\tau)$$

Also, by definition,

$$K(o) = \langle f(t) \, . \, f^*(t)\rangle = \langle |f(t)|^2 \rangle$$

The Wiener–Khinchin theorem is as follows: 'the autocorrelation function of a random function and the power density spectrum of the random function are related to one another by Fourier cosine transforms.' It is assumed that the autocorrelation function exists and that it is common to all member functions of the ensemble of representations of $f(t)$.

There are many proofs of the theorem; one which depends on the use of generalized functions is given below.

The complex fourier transform of $f(t)$ is

$$f(t) = \int_{-\infty}^{+\infty} f(\omega) \exp(\mathrm{j}\omega t) \, \mathrm{d}\omega \qquad \text{A5.1}$$

with mate $\quad f(\omega) = \dfrac{1}{2\pi} \displaystyle\int_{-\infty}^{+\infty} f(t) \exp(-\mathrm{j}\omega t) \, \mathrm{d}t$

Then $\quad f^*(\omega') = \dfrac{1}{2\pi} \displaystyle\int f^*(t') \exp(\mathrm{j}\omega' t') \, \mathrm{d}t'$

so that

$$\langle f(\omega) \, . \, f^*(\omega')\rangle = \frac{1}{4\pi^2} \iint \langle f(t) \, . \, f^*(t')\rangle \exp[\mathrm{j}(\omega' t' - \omega t)] \, \mathrm{d}t' \, \mathrm{d}t \qquad \text{A5.2}$$

Let $t = t' + \tau$ so that the r.h.s. becomes

$$\frac{1}{4\pi^2}\iint K(\tau) \exp[j(\omega' t' - \omega t')] \exp(-j\omega\tau)\, dt'\, d\tau$$

$$= \frac{1}{2\pi}\int K(\tau) \exp(-j\omega\tau)\, d\tau \,.\, \frac{1}{2\pi}\int \exp[j(\omega' t' - \omega t')]\, dt'$$

$$= \frac{1}{2\pi}\int K(\tau) \exp(-j\omega\tau)\, \delta(\omega' - \omega)\, d\tau \qquad \text{A5.3}$$

From equation A5.1 we also find

$$\langle f^*(t) \,.\, f(t)\rangle = \iint \exp[j(\omega t - \omega' t)]\, \langle f(\omega) \,.\, f^*(\omega')\rangle\, d\omega\, d\omega'$$

or, inserting equation A5.3,

$$= \frac{1}{2\pi}\iiint d\omega\, d\omega'\, d\tau\, K(\tau) \exp(-j\omega\tau)\, \delta(\omega' - \omega) \exp[j(\omega t - \omega' t)]$$

$$= \frac{1}{2\pi}\iint d\omega\, d\tau\, K(\tau) \exp(-j\omega\tau)$$

$$= \langle |f(t)|^2\rangle \qquad \text{A5.4}$$

But, by Parseval's theorem,

$$\int_{-\infty}^{+\infty} \langle |f(t)|^2\rangle\, dt = 2\pi \int_{-\infty}^{+\infty} \langle |f(\omega)|^2\rangle\, d\omega \qquad \text{A5.5}$$

and the *energy density spectrum* is defined as

$$\Phi(\omega) = 2\pi\, \langle |f(\omega)|^2\rangle$$

therefore $\displaystyle\int \langle |f(t)|^2\rangle\, dt = \int \Phi(\omega)\, d\omega$ A5.6

Comparing equations A5.6 and A5.7, we see that

$$\Phi(\omega) = \frac{1}{2\pi}\int_{-\infty}^{+\infty} d\tau \,.\, K(\tau) \exp(-j\omega\tau) \qquad \text{A5.7}$$

and, by Fourier inversion,

$$K(\tau) = \int_{-\infty}^{+\infty} d\omega\, \Phi(\omega) \exp(-j\omega\tau) \qquad \text{A5.8}$$

thus the theorem is proved.

In going from equationsA5.3 to equation A5.4 we could have taken advantage of the fact that $K(r)$ is real and we could have written

$$2K(\tau) = \langle f(t)\, f^*(t)\, f(t')\rangle$$

The result in equation A5.7 would then have been

$$\Phi(\omega) = \frac{1}{2\pi} \int \mathrm{d}\tau \frac{K(\tau)}{2} \{\exp(-\mathrm{j}\omega\tau) + \exp(\mathrm{j}\omega\tau)\}$$

$$= \frac{1}{2\pi} \int \mathrm{d}\tau\, K(\tau) \cos \omega\tau \qquad \text{A5.9}$$

and $$K(\tau) = \int \mathrm{d}\omega\, \Phi(\omega) \cos \omega\tau \qquad \text{A5.10}$$

which are often used. Accurate calculations involving complex conjugates have to be done carefully and involve cumbersome expressions, so we have abbreviated the main derivation here.

Appendix 6. Two-particle Collisions

It is assumed that the particles can be treated as rigid spheres with position vectors r_1, r_2 masses m_1, m_2 and velocities v_1, v_2. Therefore, vibrational and rotational modes are ignored as, likewise, is the spin, if any. The momentum and kinetic energy are conserved, so the basic conservation principles are

$$p_1 + p_2 = p_3 + p_4 = P = \text{constant} \quad \text{A6.1}$$

$$\text{or } v_1 + v_2 = v_3 + v_4 = P/m \quad \text{A6.2}$$

$$v_1^2 + v_2^2 = v_3^2 + v_4^2 \quad \text{A6.3}$$

where subscript 3 and 4 relate to affairs after collision.

We introduce the relative velocity $u = v_1 - v_2$, the relative radius vector $R = r_1 - r_2$, and the (constant) velocity of the centre of mass c

$$\text{where} \quad c = \frac{P}{m_1 + m_2} = \frac{m_1 v_1 + m_2 v_2}{m_1 + m_2} = \frac{d(r_c)}{dt} \quad \text{A6.4}$$

and r_c is the coordinate of the centre of mass

$$r_c = \frac{m_1 r_1 + m_2 r_2}{m_1 + m_2} \quad \text{A6.5}$$

Finally, we write $\mu = \dfrac{m_1 m_2}{m_1 + m_2}$ A6.6

Then, for equal masses $\mu = \frac{1}{2}$; while for $m_1 \gg m_2$, $\mu \to m_2$, i.e. the *smaller* mass.

In terms of c and U we can write

$$v_1 = c + \frac{\mu}{m_1} U \quad \text{A6.7}$$

$$v_2 = c - \frac{\mu}{m_2} U \quad \text{A6.8}$$

In the collision, v_1 changes to v_3, v_2 to v_4; but, since energy is conserved.

$$|U_{1,2}| = |U_{3,4}|$$

Using the momentum equations and equations A6.7 and A6.8, we can construct the momentum diagram of Fig. A6.1. The result of the collision is merely that the relative velocity vector has been turned through the angle θ, but z is not necessarily in the same plane as X, Y, W.

The trajectories are most easily visualized in the centre of mass frame, i.e. in a reference frame whose origin *always* coincides with the centre of mass. Fig. A6.2

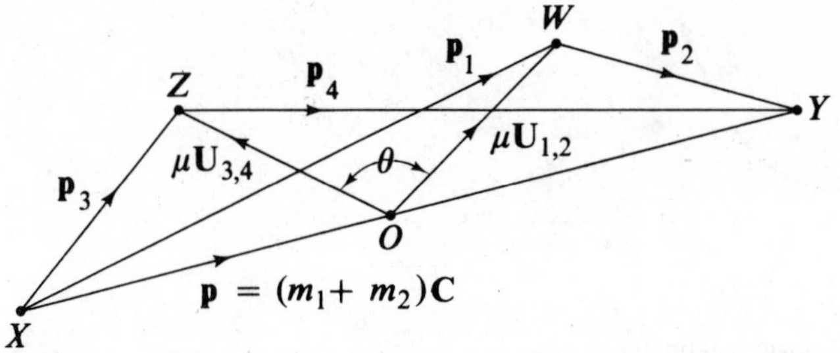

Fig. A.6.1 The momentum diagram for elastic collisions when $m_1 \approx m_2$. $XY = p_1 + p_2 = p_3 + p_4$. $XO = m_1 c$, $OY = m_2 c$. The diagram is drawn in a plane which may cut the spherical surface at any angle

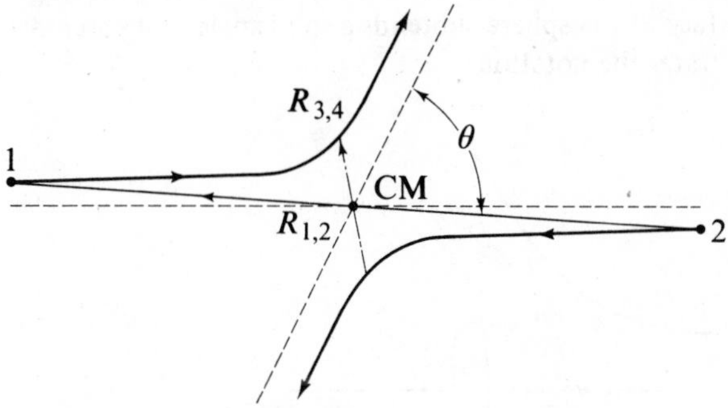

Fig. A.6.2 Notation for collision in centre of mass frame

shows the collision in this frame. Let $r_1{}^*$, $r_2{}^*$ be the position vectors relative to the c.m., then we have

$$m_1 r_1{}^* = -m_2 r_2{}^* \qquad \text{A6.9}$$

because the vectors are always oppositely directed, and

$$r_1{}^* = r_1 - r_c = \frac{\mu}{m_1} R \qquad \text{A6.10}$$

using equation A6.5. Then

$$r_2{}^* = r_2 - r_c = \frac{\mu}{m_2} R \qquad \text{A6.11}$$

In the new frame the momentum of particle 1 is $\mathrm{d}(m_1 r_1{}^*)/\mathrm{d}t = \mu\, \mathrm{d}R/\mathrm{d}t$, or the force is $\mu\, \mathrm{d}^2R/\mathrm{d}t^2$. We may then consider the c.m. to be *at rest* and matters are described by the one-particle motion of Fig. A6.3; in other words, the collision merely changes the direction of a particle which has been scattered by a fixed scatterer at the centre of mass of the system. If we consider a flow of particles, of the same mass but different velocities, incident on the scattering centre, they will be turned through different angles θ, depending on the velocity and on the so-called impact parameter marked b on Fig. A6.3. Those particles within a small velocity range and small range of b will be scattered into a small element of the surface of a sphere,

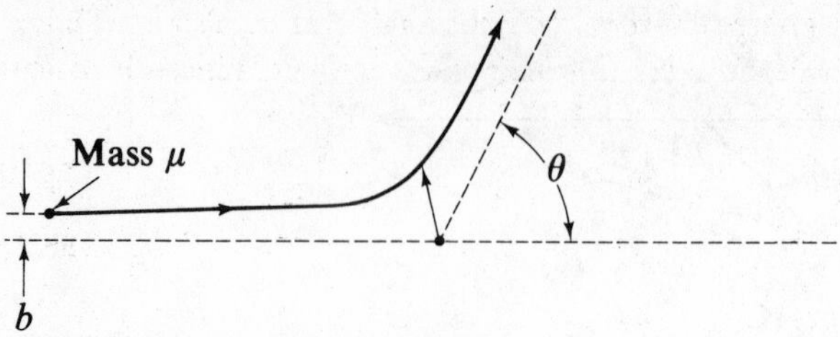

Fig. A.6.3 Definition of impact parameter.

since the angle θ can be in any plane. If we can calculate the range $d\theta$ we can calculate the area of the surface element and thus the probability of scattering into this element, since the whole surface of the sphere subtends a solid angle of 4π steradians at the centre. Fig. A6.4 illustrates the notation.

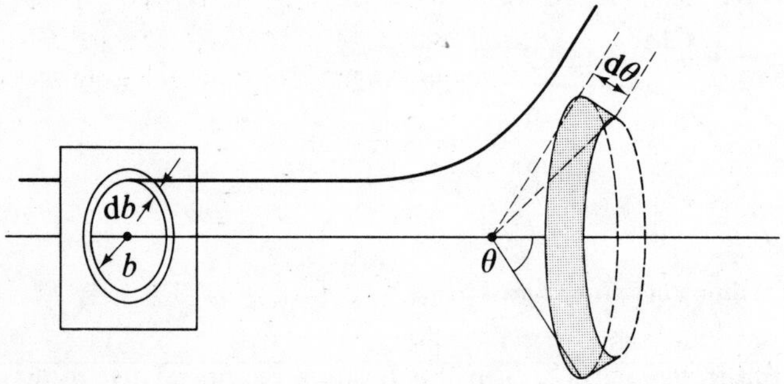

Fig. A.6.4 Notation for scattering of a particulate beam by a target particle distribution.

We imagine that a flux of particles of number density n and velocity v_0 is incident on the scattering centres. The scattering, or target, centres are of density N_T. The flux of bombarding particles is $J_0 = nv_0$ or, when we come to charged particles, is a current density nqv_0. Then, the number of particles scattered into the solid angle $d\Omega$ is

$$dn = J_0 \,.\, \sigma(\theta)\, d\Omega$$

$$= J_0 \,.\, \sigma(\theta)\, 2\pi \sin\theta d\theta$$

$$= J_0 \,.\, d\sigma$$

where $\sigma(\theta)$ = probability of scattering through angle θ
and $\quad d\sigma = \sigma(\theta)\,.\,2\pi \sin\theta.d\theta$

$$\sigma = \int_{\text{all angles}} d\sigma$$

But, since the scattered particles all have the same velocity, or energy, all the particles in db at b will be scattered through the same angle and we also must have dn= $-J_0 \,.\, 2\pi b$ db. The sign is negative because an increase in b diminishes the angle θ.

Comparing the expressions,

$$\sigma(\theta) = \frac{-b\mathrm{d}b}{\sin\theta\mathrm{d}\theta} = \frac{b}{\sin\theta}\left|\frac{\mathrm{d}b}{\mathrm{d}\theta}\right| \qquad \text{A6.12}$$

We often wish to know how many particles are scattered through angles greater than some fixed angle θ_1. To find this, the limits on b are zero and $b(\theta_1)$, the impact parameter for scattering through θ_1; and for θ are θ_1 and π. The result is

$$\tfrac{1}{2}b^2(\theta_1) = \int_{\theta_1}^{\pi} \sigma(\theta)\sin\theta\mathrm{d}\theta \qquad \text{A6.13}$$

It now remains to find the relationship between E, b and θ for all particular cases. The standard analysis, given in all textbooks on mechanics, e.g. reference 3, yields the following result for the angle between the asymptotes to the trajectory before and after collision, which we call ϕ. From Fig. A6.3, $\phi = \pi - \theta$. The required result is

$$\phi_0 = 2\int_{r_0^*}^{\infty} \frac{\mathrm{d}r^*}{r^{*2}\sqrt{\{(1/b^2)[1 - U(r_1^*)/E] - 1/r^{*2}\}}}$$

where r_0^* = distance of closest approach

$U(r_1^*)$ = potential of the force between particles

The closest approach r_0^* is determined by the equation

$$E = U(r_0^*) + b^2E/r_0^{*2}$$

For a *coulomb* collision,

$$U = \frac{|q_1q_2|}{4\pi\epsilon_0 r}$$

Some rather tedious algebra results in

$$b = \frac{|q_1q_2|}{8\pi\epsilon_0 E}\tan(\phi/2) = \frac{|q_1q_2|\cot(\theta/2)}{4\pi\epsilon_0 m v_0^2} \qquad \text{A6.14}$$

Putting this into equation A6.13 results in

$$\sigma(\theta, v_0) = \frac{1}{4m^2}\frac{|q_1q_2|^2}{(4\pi\epsilon_0)^2}\cdot\frac{1}{v_0^4\sin^4(\theta/2)} \qquad \text{A6.15}$$

Equation A6.15 is the celebrated Rutherford scattering formula, deduced in connection with his early work on the scattering of α particles. It will be much used in the discussion of scattering in plasmas. At the moment we merely note that it is independent of the sign of the charges and is inversely proportional to v_0^4 and to $\sin^4(\theta/2)$.

Scattering cross-sections have been worked out for many special cases, and the reader will find details in specialized texts on scattering theory in several branches of physics. For example, Chapman and Cowling[4] give several results relating to potentials used in gas theory.

The scattering cross-section, as defined above, is a measurable quantity, since it can be established by measurement of dn as a function of θ.

The reader should note that the above results for Rutherford scattering are written for the scattering of a light particle (an electron, for instance) by a very heavy particle, so that the reduced mass is unity. If μ takes on other values, equation A6.14 must be written as

$$b = \frac{|q_1 q_2| \cot(\theta/2)}{\mu\, 4\pi\epsilon_0 v_0^{\,2}}$$

We shall need the approximation, correct for small-angle scattering, that

$$\theta \approx \frac{2}{\mu b} \frac{|q_1 q_2|}{4\pi\epsilon_0 v_0^{\,2}}$$

In the main text we shall require expressions for the energy and momentum changes experienced by one particle in a coulomb collision. Fig. A6.5 shows the

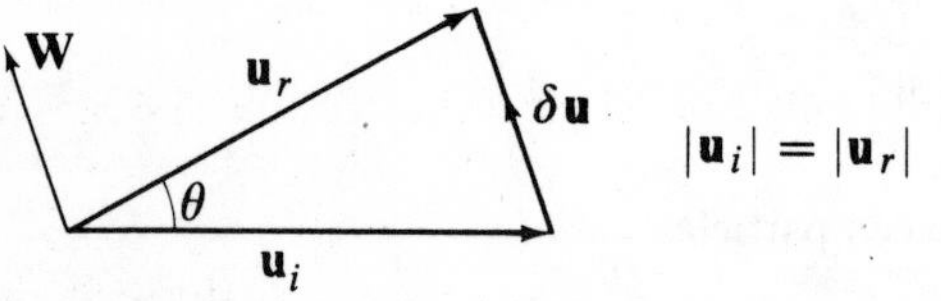

Fig. A.6.5 Vector diagram for reduced velocities during collision.

vector diagram for the collision as seen by one particle in the c.m. frame. Only the *direction* of $\boldsymbol{U}$ changes. $\boldsymbol{W}$ is a unit vector through the origin, directed normally to the plane in which the collision occurs. In an isotropic scatterer, there is no preferred direction for $\boldsymbol{W}$ and, on averaging, terms containing it drop out.

Since the magnitude of $\boldsymbol{U}$ is constant,

$$\delta(\boldsymbol{U})^2 = 2\boldsymbol{U}\delta\boldsymbol{U} + (\delta\boldsymbol{U})^2 = 0 \qquad \text{A6.16}$$

Considering the first momentum change, $\delta\boldsymbol{p}_1 = -\delta\boldsymbol{p}_2$, by conservation. From equation A6.7, $\delta\boldsymbol{p}_1 = \mu\delta\boldsymbol{U}$

Taking the components of $\delta\boldsymbol{U}$ from Fig. A6.5,

$$\delta p_{1=} = m_1 \delta U_{=} = \mu U(\cos\theta - 1) = -2\mu U \sin^2(\theta/2) \qquad \text{A6.17}$$

$$\delta p_{1_\perp} = m_1 \delta U_\perp = \mu U \sin\theta \qquad \text{A6.18}$$

Returning to the vector form,

$$\begin{aligned}\delta\boldsymbol{p}_1 &= \mu[(\boldsymbol{W} \times \boldsymbol{U}) \sin\theta - 2\boldsymbol{U} \sin^2(\theta/2)] \qquad \text{A6.19}\\ &= -\delta\boldsymbol{p}_2\end{aligned}$$

We can now obtain the energy change. If we square equation A6.7 and multiply by $m/2$, we find the energy. Varying this, remembering equation A6.16, we find

$$\delta E = \mu(\boldsymbol{C} \cdot \delta \boldsymbol{U})$$
$$= \mu[\boldsymbol{C} \cdot (\boldsymbol{W} \times \boldsymbol{U}) \sin\theta - 2\sin^2(\theta/2)(\boldsymbol{C} \cdot \boldsymbol{U})] \qquad \text{A6.20}$$

The Boltzmann collision integral

Now that we have studied the dynamics of the binary collision, we can derive the Boltzmann form of the collision integral. The following assumptions are implicit in the derivation.

1. The collision cross-sections are independent of any external forces which may act.
2. The gas, or other system, is such that only binary collisions need be considered. In fact, we cannot write a theory in which more complex interactions are included.
3. The theory as set up applies for times which are much longer than the duration of a collision and for spatial dimensions that are large compared with the range of the interaction force. Concretely, the distribution function is invariant over these times and distances.
4. Particles entering a collision have no memory of previous collisions or, more technically, their initial motions are uncorrelated.

Of these assumptions, one must be doubtful about the validity of 3 and 4, and a good deal of work, to which we make no further reference, has been done to improve the derivation.

The basic calculation we must make is the number of particles in a velocity range d**v** at **v** which are scattered *out* of an element of phase space in unit time and, conversely, the number of particles which are scattered *into* the same element. Let $f(\boldsymbol{p}, \boldsymbol{r}, t)$ represent the distribution function (we will work in momentum). Then, the number of particles in $\mathrm{d}\Gamma$ is

$$f(\boldsymbol{p}, \boldsymbol{r}, t)\, \mathrm{d}\Gamma = f(\boldsymbol{p}, \boldsymbol{r}, t)\, \mathrm{d}\boldsymbol{p}\, \mathrm{d}V$$

The scatterers, velocity $\mathbf{v}_1$, have a distribution function $f(\boldsymbol{p}_1, \boldsymbol{r}, t)$ but the number taking part in the collision is *not* $f\mathrm{d}\Gamma$ but rather equals the number of particles inside a cylindrical volume whose section is the differential scattering cross-section $\mathrm{d}\sigma$ (defined by equation 7.8) and whose length is the relative velocity C (to avoid confusion with v). This number is

$$C\mathrm{d}\sigma \,.\, f(\boldsymbol{p}_1, \boldsymbol{r}, t)\, \mathrm{d}\boldsymbol{p}_1$$

Thus, the number of collisions leading from $\mathbf{v}$ to $\mathbf{v}_1$ is

$$f(\boldsymbol{p}, \boldsymbol{r}, t)\, \mathrm{d}\boldsymbol{p}\, \mathrm{d}V \,.\, C\mathrm{d}\sigma \,.\, f(\boldsymbol{p}_1, \boldsymbol{r}, t)\, \mathrm{d}\boldsymbol{p}_1$$

or, introducing the value of $\mathrm{d}\sigma$,

$$C\sigma(c,\theta)\, f(\boldsymbol{p},\boldsymbol{r},t)\, f(\boldsymbol{p}_1,\boldsymbol{r},t)\, \mathrm{d}\Omega\, \mathrm{d}\boldsymbol{p}\, \mathrm{d}V\, \mathrm{d}\boldsymbol{p}_1 \qquad \text{A6.21}$$

We recall that equation A6.21 gives the number of particles scattered *out* of the element $\mathrm{d}p\mathrm{d}V$ in the particular class $\mathbf{v} \to \mathbf{v}_1$. Call this $ar \,.\, \mathrm{d}\boldsymbol{p}\mathrm{d}V$; then

$$a = \iint C\sigma \,.\, f(\boldsymbol{p}, r, t) f(\boldsymbol{p}_1, r, t)\, \mathrm{d}\Omega\, \mathrm{d}\boldsymbol{p}_1 \qquad \text{A6.22}$$

We can easily see that, by the same reasoning, the number of particles scattered *into* the phase-space element $\mathrm{d}\boldsymbol{p}_2\, \mathrm{d}V$ is

$$C\sigma' f(\boldsymbol{p}_2, r, t) f(\boldsymbol{p}_3, r, t)\, \mathrm{d}\Omega\, \mathrm{d}v\, \mathrm{d}\boldsymbol{p}_2\, \mathrm{d}\boldsymbol{p}_3$$

Here, σ is the differential cross-section for $v_2 \to v_3$.

Next, we have to specialize to the case

$$|C| = |\mathbf{v} - \mathbf{v}| = |\mathbf{v}_2 - \mathbf{v}_3|$$

Clearly, $\sigma = \sigma'$ when this is so; but we still have to find the transition law $\mathrm{d}\boldsymbol{p}\, \mathrm{d}p_1 \to \mathrm{d}\boldsymbol{p}_2\, \mathrm{d}\boldsymbol{p}_3$. Perhaps the easiest way to do this is in the components, remembering equations A6.7 and A6.8 and that m_1 now equals m_2 and therefore $\mu = m_1/2$.

Forming the Jacobian of the transformation we get, from equations A6.7 and A6.8,

$$\mathrm{d}v_{11}\, \mathrm{d}v_{21} = \begin{vmatrix} 1 & \mu/m_1 \\ 1 & -\mu/m_2 \end{vmatrix} \mathrm{d}c_1\, \mathrm{d}C_1$$

$$= -\mathrm{d}c_1\, \mathrm{d}C_1$$

or, returning to the vector form,

$$\mathrm{d}\mathbf{v}_1\, \mathrm{d}\mathbf{v}_1 = \mathrm{d}c\, \mathrm{d}C$$

Since exactly the same holds for $\mathrm{d}v_2\, \mathrm{d}v_3$, we find

$$\mathrm{d}\boldsymbol{p}\, \mathrm{d}\boldsymbol{p}_1 = \mathrm{d}\boldsymbol{p}_2\, \mathrm{d}\boldsymbol{p}_3$$

Using this and replacing σ' by σ we find

$$b = \iint C\sigma f(\boldsymbol{p}_2, r, t_1) f(\boldsymbol{p}_3, r, t)\, \mathrm{d}\Omega\, \mathrm{d}\boldsymbol{p}_1 \qquad \text{A6.23}$$

Thus, the net gain or loss of particles, which is $b \sim a$, is just

$$\iint C\sigma[f(\boldsymbol{p}_2) f(\boldsymbol{p}_3) - f(\boldsymbol{p}) \,.\, f(\boldsymbol{p}_1)]\, \mathrm{d}\Omega\, \mathrm{d}\boldsymbol{p}_1 \qquad \text{A6.24}$$

Since σ is a function of the relative velocity, the collision integral is non-linear and it is extremely difficult to solve the Boltzmann equation with equation A6.19 on the r.h.s.

In the main text, equation A6.19 was slightly modified by identifying $\boldsymbol{p}, \boldsymbol{p}_1, \boldsymbol{p}_2, \boldsymbol{p}_3$ $\boldsymbol{p}_1, \boldsymbol{p}_2, \boldsymbol{p}_3, \boldsymbol{p}_4$ to make it look more symmetrical.

Appendix 7. Physical Definitions of the Moments

This appendix discusses the relationship between the moments of a distribution function in velocity and the various quantities which depend on them. These quantities were defined in fluid and gas dynamics before statistical mechanics was fully established.

We assume that the distribution function is normalized to N particles, i.e. $N = \int f \, d\mathrm{v}$, also, f is a function of v which decreases rapidly as $\mathrm{v} \to \pm\infty$. Otherwise f need not be symmetrical and is certainly not assumed to be Maxwellian. We shall be discussing quantities including the number density, the temperature, the pressure tensor and the energy flux, and derived quantities such as viscosity and compressibility. In charged-particle flows, plasmas for example, the same concepts are used but the terminology is different and we have to include electrons and at least one species of ions. This special case is discussed in every introduction to plasma physics, but all we need to know here is that equations can be set up to describe the separate flow properties of electrons and ions. A last general point is that there are two conventions for specifying particle velocities. In one convention, the random velocity $\mathrm{v} - \langle \mathrm{v} \rangle$ is used; in the other, which is useful when the fluid is in motion as a whole, $\boldsymbol{V} - \boldsymbol{U}$ is used, where $\boldsymbol{U}$ is the velocity of the centre of mass. The formulae appear somewhat differently in the two frames.

For convenience, we repeat the basic transport equation

$$\frac{\partial}{\partial t}(N\langle\phi\rangle) + \frac{\partial}{\partial r}(N\langle\phi \mathrm{v}\rangle) - \frac{Ne}{m}[E + \nabla.\mathrm{x}\, B]\left\langle\frac{d\phi}{d\mathrm{v}}\right\rangle = \left(\frac{\partial}{\partial t}N\langle\phi\rangle\right)_{\mathrm{c}} \qquad \text{A7.1}$$

where we can think of the E and B contributions as those of generalized velocity-independent and velocity-dependent forces, rather than as electric and magnetic forces. We let ϕ successively assume the values 1, v, $\frac{1}{2}mv^2$. However, we need not, for the moment, use the Boltzmann form for the collision term, which for these values of ϕ automatically becomes zero. We can be a little more general and assume, for example, that N is not necessarily conserved; for example, this would be the case if we considered ionization.

1. Putting $\phi = 1$ in equation A7.1, we find

$$\frac{\partial N}{\partial t} + \frac{\partial}{\partial r} \cdot (N\langle \mathrm{v}\rangle) = \left(\frac{\partial N}{\partial t}\right)_{\mathrm{c}} \qquad \text{A7.2}$$

or, in more easily recognizable form,

$$\frac{\partial \rho}{\partial t} + \nabla \cdot (\rho\langle \mathrm{v}\rangle) = \left(\frac{\partial \rho}{\partial t}\right)_{\mathrm{c}} \qquad \text{A7.3}$$

where $\rho = Nm$, the ordinary density.

Equation A7.3 is the standard form of the equation of continuity, which is very frequently written with zero r.h.s.

2. $\phi_i = mv_i$ gives the component equation

$$\frac{\partial}{\partial t}(Nmv_i) + \nabla \cdot (Nm\langle \mathbf{v}.v_i\rangle) - NeE_i - Ne(\nabla \times B)_i = \frac{\partial}{\partial t}(Nmv_i)_c \quad \text{A7.4}$$

3. $\phi = \frac{1}{2}mv^2$

$$\frac{\partial}{\partial t}(N\langle mv^2\rangle) + \{\nabla\, N(\langle mv^2\,\mathbf{v}\rangle)\} - NeE\,.\,\mathbf{v} = \frac{\partial}{\partial t}\{N\langle mv^2\rangle\}_c \quad \text{A7.5}$$

We now have to introduce the pressure and heat tensors. For the *c.m. frame*, the pressure tensor is defined by

$$p_{ij} = Nm\langle c_i c_j\rangle = m\int c_i c_j f \mathrm{dv} \quad \text{A7.6}$$

where $c = \mathbf{v} - U$.

From equation A7.6, $p_{ij} = p_{ji}$, which means that the pressure tensor is symmetrical and has six independent components.

In the same frame, the heat-flow tensor is

$$g_{ijk} = Nm(c_i c_j c_k) = m\int c_i c_j c_k\, f\, \mathrm{dv} \quad \text{A7.7}$$

a third-order tensor, with basically 16 components. This tensor is usually contracted to a second-order tensor, a *vector* called the 'heat-flow' vector and defined as

$$Q_i = \tfrac{1}{2}q_{ijj} = \frac{m}{2}\int c_i c^2 f\, \mathrm{dv} \quad \text{A7.8}$$

Q has a simple physical meaning – it gives the rate of energy flow through a surface in the c.m. frame.

As a last definition, we recall that the temperature is defined by

$$\tfrac{3}{2}NkT = \frac{m}{2}\int c^2 f \mathrm{dv} \quad \text{A7.9}$$

Comparing equations A7.7 and A7.9,

$$T = \frac{p_{ii}}{3Nk} \quad \text{A7.10}$$

the pressure terms being those on the diagonal of the tensor.

Using these definitions, we can bring equations A7.4 and A7.5 to their more usual forms by eliminating $\langle \mathbf{v}\,.\,v_i\rangle$, $\langle mv^2\,.\,\mathbf{v}\rangle$ etc. For example,

$$\langle v_i v_j\rangle = p_{ij}/Nm + v_1\ U_j + v_j U_i - U_i U_j \quad \text{A7.11}$$

$$\langle v^2 v_i\rangle = \frac{Q_i}{m} + \frac{p_{ij}}{Nm} + \frac{2p_{ij}}{Nm} + 2v_j U_j U_i + v_i U_j U_i - 2U_i U_j U_k \quad \text{A7.12}$$

It is now straightforward to show that

$$\rho \frac{\mathrm{D}\boldsymbol{U}}{\mathrm{D}t} + \nabla \,.\, \boldsymbol{P} - e\boldsymbol{E} - Ne(\boldsymbol{V} \times \boldsymbol{B}) = 0 \qquad \text{A7.13}$$

where $\mathrm{D}/\mathrm{D}t$ is the convective derivative:

$$\mathrm{D}/\mathrm{D}t = \partial/\partial t + \nabla \,.\, \boldsymbol{U} \qquad \text{A7.14}$$

$\boldsymbol{P}$ is the dyadic form of the pressure tensor – that is $\boldsymbol{P} = \boldsymbol{a}_i \,.\, p_{ij} \cdot \boldsymbol{a}_j$, where $\boldsymbol{a}_i, \boldsymbol{a}_j$ are *unit* vectors along the i, j axes – and we have now dropped the collision term. The energy equation, similarly, is now expressible either in pressure or temperature as

$$\frac{\partial}{\partial t}\left(\frac{p_{ii}}{2} + \tfrac{1}{2}\rho U^2\right) + \nabla \,.\, [(\tfrac{1}{2}p_{ii} + \tfrac{1}{2}\rho U^2)\,\boldsymbol{U} + \boldsymbol{Q} + \boldsymbol{P} \,.\, \boldsymbol{U}] - Nev \,.\, \boldsymbol{E} = 0 \qquad \text{A7.15}$$

or $$\frac{\partial}{\partial t}(\tfrac{3}{2}NkT + \tfrac{1}{2}\rho U^2) + \nabla \,.\, [(\tfrac{3}{2}NkT + \tfrac{1}{2}\rho U^2)\,\boldsymbol{U} + \boldsymbol{Q} + \boldsymbol{P} \,.\, \boldsymbol{U}] - Nev \,.\, \boldsymbol{E} = 0 \qquad \text{A7.16}$$

If one is speaking of real charged-particle flows, the electric current density $J = Nev$ is always used. The meaning of these equations, which is somewhat different in different areas of interest, is exhibited by integration over a small volume and identification of the fluxes through the surface, the work done at the surface, and the power dissipation.

If one is working with the random velocity $\boldsymbol{C} = \mathbf{v} - \langle \mathbf{v} \rangle$, equations A7.11 and A7.12 are somewhat simplified because $\langle \boldsymbol{C} \rangle = 0$. Then

$$\langle v_i \,.\, v_j \rangle = P_{ij}/Nm + v_i \,.\, v_j \qquad \text{A7.17}$$

The momentum equation, equation A7.12, is now

$$\rho \frac{\mathrm{D}}{\mathrm{D}t} \langle \mathbf{v} \rangle + \nabla \,.\, \boldsymbol{p} - Ne(\langle v \rangle \times B) - e\boldsymbol{E} = 0 \qquad \text{A7.18}$$

and equation A7.15 is modified to

$$\tfrac{3}{2}NkT\left[\frac{\mathrm{D}}{\mathrm{D}t}\langle \mathbf{v} \rangle + \nabla \,.\, \langle \mathbf{v} \rangle\right] + \boldsymbol{p} \colon \nabla \langle \mathbf{v} \rangle + \nabla \,.\, \boldsymbol{q} = 0 \qquad \text{A7.19}$$

where $\boldsymbol{p}$ and $\boldsymbol{q}$ are defined in terms of the modified velocity.

Looking back over these derivations, it is seen that N, $\langle v \rangle$, T (obviously) and $\boldsymbol{P}$, $\boldsymbol{Q}$ by definition are all moments of the velocity distribution first discussed in Chapter 1. Therefore, if we were in the happy position of knowing *all* the moments, we should be able to determine f exactly. Since this is impossible, we must use those moments we *do* know to find an approximation to f. This is the idea behind the moment methods for solving the Boltzmann transport equation. These methods have been systematized, the work of Grad being particularly significant in this respect. The approximation technique has to be chosen with physical reality in mind – for example, we soon find we cannot merely ignore, say, $\boldsymbol{q}$; instead we can determine that, possibly under special circumstances, it is allowable to ignore some of the off-diagonal components p_{ij} and q_{ij}, with consequent reduction in the number of moments which must be included. A wrong choice of 'ignorables' will ensure that the equations do *not* form the closed set which is necessary if they are to be soluble. A well-known approximation puts $q_{ijk} = 0$ and takes $p_{ij} \equiv p_{ii}$. This is called the eight-moment approximation" and is widely used in magnetohydrodynamics; it is not useful in fluid mechanics since neglecting q is equivalent to leaving out the viscosity.

Appendix 8. The Evaluation of $\langle \Delta v \,.\, \Delta v \rangle$

In general, $\langle \Delta v \,.\, \Delta v \rangle$ is a second-order tensor, but we shall shortly see that it is diagonal and thus has only three finite components. Furthermore, the longitudinal component will turn out to be smaller than the (equal) remaining components, by approximately L, i.e. by a factor of more than 10 in most plasmas; therefore it can often be neglected.

We work in the c.m. frame before going back to the laboratory frame, and recall from equation A6.7 that $\Delta \mathbf{v} = (\mu/m_1)\, \Delta \boldsymbol{U}$. The Cartesian coordinate system is chosen so that it has right-handed unit vectors $\boldsymbol{a}_x$, $\boldsymbol{a}_y$, $\boldsymbol{a}_z$, and the incident particle velocity

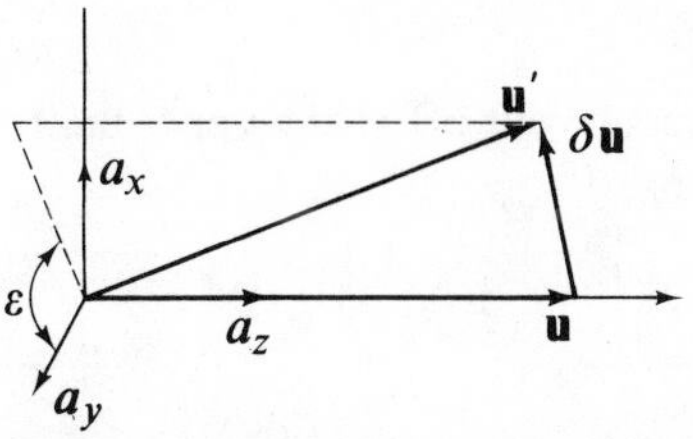

Fig. A.8.1 Collison geometry in a right-handed Cartesian system. Particle is incident along a_z.

is directed along $\boldsymbol{a}_z$. The geometry of the collision is redrawn in Fig. A8.1, showing the projection of the final relative velocity on the 2, 3 plane where it makes angle ϵ with the 3 axis. Then, slightly modifying the notation of equation A6.19, we find

$$\Delta \boldsymbol{U} = 2U \frac{\mu}{m} \sin(\theta/2)\, [-\boldsymbol{a}_z \sin(\theta/2) + \boldsymbol{a}_x \cos(\theta/2) \cos\epsilon + \boldsymbol{a}_y \cos(\theta/2) \sin\epsilon] \quad \text{A8.1}$$

When this is squared and integrated over all values of ϵ, only the terms in $\boldsymbol{a}_z \,.\, \boldsymbol{a}_z$, $\boldsymbol{a}_x \,.\, \boldsymbol{a}_x$, $\boldsymbol{a}_y \,.\, \boldsymbol{a}_y$ survive, so the tensor is diagonal. Moreover, the two transverse components are equal in magnitude. Thus, the tensor $\Delta \boldsymbol{U} \,.\, \Delta \boldsymbol{U}$ can be written

$$\Delta \boldsymbol{U} \,.\, \Delta \boldsymbol{U} = \begin{bmatrix} \alpha & 0 & 0 \\ 0 & \alpha & 0 \\ 0 & 0 & \beta \end{bmatrix} \quad \text{A8.2}$$

The values of α and β can be derived using the Rutherford scattering formula, but it proves convenient to use the expression

$$\sin(\theta/2) = \frac{b_\perp}{\sqrt{(b_\perp^2 + b^2)}} \quad \text{A8.3}$$

to eliminate the angular function. This is a generalization of equation 8.34. Then, using equations A6.17 and A6.18, we find

$$\begin{matrix}\Delta U_x \\ \Delta U_y\end{matrix} = 2U\frac{\mu}{m}\sin(\theta/2)\cos(\theta/2)\begin{matrix}\sin\epsilon \\ \cos\epsilon\end{matrix} = 2U\frac{\mu}{m}\frac{bb_\perp}{b_\perp^2+b^2}\begin{matrix}\sin\epsilon \\ \cos\epsilon\end{matrix} \qquad \text{A8.4}$$

$$\Delta U_z = 2U\frac{\mu}{m}\frac{b_\perp^2}{b_\perp^2+b^2} \qquad \text{A8.5}$$

The number of scattering events due to incident particles with impact parameters in db at b, and orientation dϕ at ϕ is $ubdbd\phi$, integrated over all b, over ϕ ranging from 0 to 2π and lastly over the distribution of the scattering particles.

Let us define our tensor so that we shall finally obtain

$$\langle \Delta \mathrm{v}\, \Delta \mathrm{v}\rangle = \int T_{ij}\, f_\alpha(\mathrm{v_s})\mathrm{d}\mathrm{v_s}$$

In the c.m. frame the components are

$$T_{xx} = T_{yy} = \pi\left(\frac{\mu}{m}\right)^2\int_0^\infty\left(\frac{2Ubb_\perp}{b_\perp^2+b^2}\right)^2 Ubdb \qquad \text{A8.6}$$

The integral in equation A8.6 is divergent and must be cut off at the upper limit λ_D as usual. If we do this, it becomes

$$\ln\left(\frac{\lambda_D^2+b_\perp^2}{b_\perp^2}\right)^{\frac{1}{2}} - \tfrac{1}{2} \approx \ln\left(\frac{\lambda_D^2+b_\perp^2}{b_\perp^2}\right)^{\frac{1}{2}}$$

$$\approx \ln\left(\frac{\lambda_D}{b_\perp}\right) = L \qquad \text{A8.7}$$

since $\lambda_D/b_\perp \gg 1$ and $\frac{1}{2}$ is negligible compared with L.

We can now return to the laboratory frame by substituting

$$b_\perp = \frac{|ee_s|}{4\pi\epsilon_0\mu U^2} \qquad \text{A8.8}$$

to obtain the final result for

$$T_{xx} = T_{yy} = \frac{4\pi\,|ee_s|^2}{(4\pi\epsilon_0 m)^2}\cdot\frac{1}{U} \qquad \text{A8.9}$$

Similarly,

$$T_{zz} = \left(\frac{\mu}{m}\right)^2\int_0^\infty\left(\frac{2Ub_\perp^2}{b_\perp^2+b^2}\right)^2 Ubdb$$

$$= 4\left(\frac{\mu}{m}\right)^2 U^3 b_\perp^4\int_0^\infty\frac{bdb}{(b_\perp^2+b^2)^2}$$

$$= 2\left(\frac{\mu}{m}\right)^2 U^3 b_\perp^2 \qquad \text{A8.10}$$

This is smaller than T_{XX}, by the factor $2\pi L$, which is much greater than unity, and therefore can be neglected.

Putting the results into vector form, we arrive at

$$\langle\Delta\mathrm{v}\,.\,\Delta\mathrm{v}\rangle = \frac{4\pi|ee_s|^2}{(4\pi\epsilon_0 m)^2}\int \frac{(a_X a_X + a_Z a_Z)}{U}\, f(\mathrm{v_s})\,\mathrm{dv_s} \qquad \text{A8.11}$$

or, in tensor notation,

$$\langle\Delta\mathrm{v}\,.\,\Delta\mathrm{v}\rangle = \frac{4\pi|ee_s|^2}{(4\pi\epsilon_0 m)^2}\int \frac{U^2\delta_{ij} - U_i U_j}{U^3}\, f(\mathrm{v_s})\,\mathrm{dv_s} \qquad \text{A8.12}$$

where δ_{ij} is the Kronecker delta symbol. Since $\boldsymbol{u}$ is the velocity-space analogue of the position vector $\boldsymbol{R}$ in ordinary space, we can use the standard vector identity

$$\frac{(U^2 I - UU)}{U^3} = \frac{\partial^2 U}{\partial\mathrm{v}\,.\,\partial\mathrm{v}}$$

or the tensor form of the same result:

$$\frac{(U^2\delta_{ij} - U_i U_j)}{U^3} = \frac{\partial^2 U}{\partial\mathrm{v_i}\,.\,\partial\mathrm{v_j}}$$

to obtain the second term of the Fokker–Planck equation in the form in which it is usually written:

$$\frac{4\pi|ee_s|^2}{(4\pi\epsilon_0 m)^2}\,.\,\tfrac{1}{2}\,.\,\frac{\partial}{\partial\mathrm{v}\,.\,\partial\mathrm{v}} : \left[f(\mathrm{v_s})\,.\,\frac{\partial^2 G(\mathrm{v})}{\partial\mathrm{v}\,.\,\partial\mathrm{v}}\right] \qquad \text{A8.13}$$

References

Chapter 1

a. Introductory:

1. Meyer, P. L., *Introductory Probability and Statistical Applications,* Reading Mass., Addison–Wesley, 2nd ed. 1970
2. Thomas, J. B., *Applied Probability and Random Processes,* New York, Wiley, 1971

b. More advanced:

3. Gnedenko, B. V., *The Theory of Probability,* New York, Chelsea Pub. Co., 4th ed. 1967
4. Renyi, A., *Probability Theory,* Amsterdam, North Holland, 1970

These books, while mathematically rigorous, still include discussion of applications.

c. Mathematical treatises:

5. Doob, J., *Stochastic Processes,* New York, Wiley, 1953
6. Loeve, M., *Probability Theory,* Princeton, Van Nostrand, 3rd ed. 1963

d. Specialized work on statistical mechanics:

7. Khinchin, A. I., *Mathematical Foundations of Statistical Mechanics,* New York, Dover, 1949
8. Khinchin, A. I., *Mathematical Foundations of Quantum Statistics,* Albany NY, Graylock, 1960

Chapter 2

1. Reif, F., Statistical Physics, *Berkeley Physics Course,* Vol. 5, New York, McGraw–Hill, 1965, pp. 24–5
2. Chapman, S., and Cowling, T. G., *Mathematical Theory of Non-uniform Gases,* Cambridge University Press, 2nd ed. 1960, p. 75
3. Levich, B. G., *Theoretical Physics,* Vol. 2, 'Statistical Physics', Amsterdam, North Holland, 1971, pp. 43–7
4. Swift, J. D., 'Kinetic Theory of Gases and Gaseous Flow', in Beck, A.H.W., (ed.) *Handbook of Vacuum Physics* Part 5, Oxford, Pergamon, 1966

Chapter 3

1. Goldstein, H., *Classical Mechanics,* Reading Mass., Addison–Wesley, 1959
2. Corben, H. C., and Stehle, P., *Classical Mechanics,* New York, Wiley, 2nd ed. 1960
3. Mercier, A., *Analytical and Canonical Formalism in Physics,* Amsterdam, North Holland, 1959. Reprinted, New York, Dover, 1963. Chapter 3, pp. 97–125. [Advanced, but clear and compact]
4. Landau, L. D., and Lifschitz, E. M., *Statistical Physics,* Oxford, Pergamon, 2nd ed. 1968
5. Schrödinger, E., *Statistical Thermodynamics,* Cambridge University Press, 1st ed. 1946, 2nd ed. 1952
6. Lindsay, R. B., *Introduction to Physical Statistics,* New York, Wiley, 1941. [Gives introduction to Darwin–Fowler method.]
7. Fowler, R. H., *Statistical Mechanics,* Cambridge University Press, 2nd ed. 1955
8. Fowler, R. H., and Guggenheim, E. A., *Statistical Thermodynamics,* Cambridge University Press, 2nd ed. 1949

9. Khinchin, A. I., *Mathematical Foundations of Statistical Mechanics,* New York, Dover, 1949
10. Khinchin, A. I., *Mathematical Foundations of Quantum Statistics,* Albany NY, Graylock, 1960
11. Huang, K., *Statistical Mechanics,* New York, Wiley, 1963
12. Morse, P. M., *Thermal Physics,* New York, 2nd ed. Benjamin, 1964

Chapter 4

Good treatments of quantum theory are given by

1. Dicke, R. H., and Wittke, J. P., *Introduction to Quantum Mechanics,* Reading Mass., Addison–Wesley, 1960
2. Park, D., *Introduction to the Quantum Theory,* New York, McGraw-Hill, 1964
3. Davydov, A. S.; tr. Ter Haar, D., *Quantum Mechanics*, Oxford, Pergamon, 1965
4. Kittel, C., *Introduction to Solid-State Physics,* New York, Wiley, 4th ed. 1971
5. Ziman, J. M., *Principles of the Theory of Solids,* Cambridge University Press, 2nd ed. 1972

Chapter 5

1–3. References 1, 2, and 3 of Chapter 4 cover the differences between fermions and bosons.
4. Reference 5 of Chapter 4 contains an introduction to superconductivity.

A more extensive treatment is given in

5. Rhoderick, E. H., and Rose–Innes, A. C., *Introduction to Superconductivity,* Oxford. Pergamon, 1969

The following is much more advanced:

6. Blatt, J. M., *Theory of Superconductivity,* New York, Academic, 1964.

Chapter 6

1. Khinchin, A. I., *Mathematical Foundations of Statistical Mechanics,* New York, Dover, 1949
2. Einstein, A., *Investigations on the Theory of the Brownian Movement,* ed. R. Fürth, New York, Dover, 1956
3. Langevin, P., *Comptes rendus,* Vol. 146, 1908, p. 530
4. Carslaw, H. S., and Jaeger, J. C., *Conduction of Heat in Solids,* Oxford University Press, 1959
5. Stratonovitch, R. L., *Topics in the Theory of Random Noise,* Vol. 1, Chapter 4, New York, Gordon and Breach, 1963

Chapter 7

1. Grad, H., 'Principles of the kinetic theory of gases' in *Handbuch der Physik,* Vol. 12, Berlin, Springer Verlag, 1958
2. Grad, H., *Commun. Pure and Applied Math.,* Vol. 2, 1949, p. 331
3. Brush, S. G., *Kinetic Theory,* Vol. 1, 1965, Vol. 2, 1966, Vol. 3, 1972, Oxford, Pergamon Press. [A collection of original papers]
4. Chapman, S., and Cowling, T. G., *The Mathematical Theory of Non-uniform Gases,* Cambridge University Press, 2nd ed. 1960, p. 67
5. Levich, B. G., *Theoretical Physics,* Vol. 2. 'Statistical Physics', Amsterdam, North Holland. 1971, pp. 155 *et seq.*
6. Heald, M. A., and Wharton, C. B., *Plasma Diagnostics with Microwaves,* New York, Wiley, 1965
7. Boyd, T. J. M., and Sanderson, J. J., *Plasma Dynamics,* London, Nelson, 1969
8. Blatt, F. J., *Physics of Electronic Conduction in Solids,* New York, McGraw–Hill, 1968
9. Kogan, M. N., *Rarefied Gas Dynamics,* New York Plenum, 1969
10 Cercignani, C., *Mathematical Methods in Kinetic Theory,* New York Plenum, 1969
11. Park, D., *Introduction to the Quantum Theory,* New York, McGraw–Hill, 1964
12. Liboff, R. L., *Introduction to the Theory of Kinetic Equations,* New York, Wiley, 1969

Chapter 8

1. Vlasov, A. A., *Zh. Eksp. Teor. Fiz.* Vol. 8, 1938, p. 291 [Russian original].
   Vlasov, A. A., *Many-Particle Theory and its Application to Plasma,* New York, Gordon and Breach, 1961
2. Landau, L., *J. Phys. USSR,* Vol. 10, 1946, p. 25
3. Bohm, D., and Gross, E. P., *Phys. Rev.* Vol. 75, 1949, pp. 1851 and 1854
4. Kaplan, W., *Operational Methods for Linear Systems,* Reading Mass., Addison–Wesley, 1962
5. Gakhov, F. D., *Boundary Value Problems,* Oxford, Pergamon, 1966
6. Ecker, G., *Theory of Fully Ionized Plasmas,* New York and London, Academic, 1972
7. Krall, N. A., and Trivelpiece, A. W., *Principles of Plasma Physics,* New York, McGraw-Hill, 1973
8. Sivukhin, D. V., 'Coulomb collisions in a fully ionized plasma', *Reviews of Plasma Physics,* Vol 4, New York, Consultants Bureau, 1966
9. Rosenbluth, M. N., MacDonald, W. M., and Judd, D. L., *Phys. Rev.,* Vol. 107, 1957, p. 1
10. Chandrasekhar, S., *Principles of Stellar Dynamics,* New York, Dover, 1960, p. 63
11. Trubnikov, B. A., 'Particle Interactions in a Fully Ionized Plasma', *Reviews of Plasma Physics,* Vol. 1, New York, Consultants Bureau, 1965
12. Shkarofsky, I. P., Johnston, T. W., and Bachynski, M. P., *Particle Kinetics of Plasmas,* Reading Mass., Addision–Wesley, 1966
13. Landau, L., *J. Phys. USSR,* Vol. 10, 1936, p. 154. *J. Exptl. Theoretical Phys. USSR,* Vol. 7, 1937, p. 203. Reprinted in Landau, L., *Collected Papers,* Oxford, Pergamon Press, 1965
14. Braginskii, S. I., *Reviews of Plasma Physics,* Vol. 1, New York, Consultants Bureau, 1965
15. Lenard, A., *Anns. Phys.* (New York), Vol. 10, 1960, p. 390
16. Balescu, R., *Phys. Fluids,* Vol. 3, 1960, p. 52
17. Bogliubov, N. N., *Studies in Statistical Mechanics,* ed. de Boer, J. H., and Uhlenbeck, G. E., Amsterdam, North Holland, 1962
18. Balescu, R., *Statistical Mechanics of Charged Particles,* New York, Interscience, 1963
19. Liboff, R. L., *Introduction to the Theory of Kinetic Equations,* New York, Wiley, 1969
20. Wu, T. Y., *Kinetic Equations of Gases and Plasmas,* Reading Mass., Addison–Wesley, 1966

Chapter 9

1. Nyquist, H., *Phys. Rev.,* Vol. 32, 1928, p. 110
2. Ming Chen Wang and Uhlenbeck, G. E., *Rev. Mod. Phys.,* Vol. 17, 1945, p. 323. Reprinted in Wax, N., (ed.), *Noise and Stochastic Processes,* New York, Dover, 1954
3. Beck, A. H. W., *Thermionic Valves,* Cambridge University Press, 1953
4. Thompson, B. J., North D. O., and Harris, W. A., 'Fluctuations in space-charge limited currents at moderately high frequencies', *RCA Review,* Jan. 1940 onwards. Reprinted in *Electron Tubes,* Vol. 1 (1935–41), Princeton NJ, RCA Laboratories Division, 1949
5. Callen, H. B., and Welton, T. A., *Phys. Rev.,* Vol. 83, 1951, p. 34
6. Landau, L. D., and Lifschitz, E. M., *Statistical Physics,* Oxford, Pergamon, 2nd ed. 1968
7. Levich, B. G., *Theoretical Physics,* Vol. 4, 'Quantum Statistics and Physical Kinetics', Amsterdam, North Holland, 1973

Chapter 10

Books and articles on information theory tend to be very easy or very difficult. Some of the easier ones are:

1. Pierce, J. R., *Symbols, Signals and Noise,* New York, Harper, 1961
2. Shannon, C. E., *Bell System Technical Journal,* Vol. 27, 1948, pp. 379 and 623
3. Shannon, C. E., and Weaver, W., *Mathematical Theory of Communication,* Urbana Ill., Univ. Illinois Press, 1949

4. Reza, F. M., *Introduction to Information Theory*, New York, McGraw–Hill, 1961

Harder:

5. Fano, F. M., *Transmission of Information,* Cambridge Mass., MIT Press, 1961
6. Brillouin, L., *Science and Information Theory,* New York, Academic, 1956. [Brillouin sets out the relationship between statistical physics and information theory in detail.]
7. Jelinek, F., *Probabilistic Information Theory,* New York, McGraw–Hill, 1968
8. Khinchin, A. I., *Mathematical Foundations of Information Theory,* New York, Dover, 1957
9. Middleton, D., *Introduction to Statistical Communication Theory,* New York, McGraw–Hill, 1960 [The use of the word 'introduction' in the title is an example of extreme over-modesty.]

Appendices

1. Levich, B. G., *Theoretical Physics,* Vol. 4, 'Quantum statistics and physical kinetics', North Holland, Amsterdam, 1973
2. Chandrasekhar, S., 'Stochastic Processes in Physics and Astronomy', *Rev. Modern Phys.,* Vol. 15, 1943, p. 1. [This paper is reproduced in the extremely useful compilation, Wax, N., *Selected Papers on Noise Stochastic Processes,* New York, Dover, 1954.]
3. Goldstein, H., *Classical Mechanics,* Reading Mass., Addison–Wesley, 1959
4. Chapman, S., and Cowling, T. G., *The Mathematical Theory of Non-Uniform Gases,* Cambridge University Press, 2nd ed. 1960

Index